# Making Math Meaningful™

# A Source Book for Teaching Math in Grades One through Five

## Second Edition

By Nettie Fabrie, Wim Gottenbos, and Jamie York

Jamie York Press

2019

*Making Math Meaningful*™

# A Source Book for Teaching Math in Grades One through Five

**Second Edition**

Text Copyright © 2009 by Nettie Fabrie, Wim Gottenbos, and Jamie York

ISBN: 978-1-938210-20-4

All rights reserved. No part of this book
may be reproduced in any form or
by any means, without permission in
writing from the publisher.

Cover Design Copyright © 2003 by Catherine Douglas.
Catherine Douglas was a student at Shining Mountain Waldorf School when she designed the cover to this book. It is an impressive example of how equiangular spirals emerge from nested octagons.

JAMIE YORK
PRESS

*Meaningful Math for Waldorf,
Public, Private, and Home Schools*

www.JamieYorkPress.com

# Table of Contents

**INTRODUCTION** ............... 1
   *What are the symptoms of our 'math illness'?* ............... 2
   *Principles of Waldorf education* ............... 5
   *Will our children be prepared?* ............... 8
   *In the math classroom* ............... 11
   *The art of teaching math* ............... 14

**FIRST GRADE MATH** ............... 23

**SECOND GRADE MATH** ............... 39

**THIRD GRADE MATH** ............... 52

**FOURTH GRADE MATH** ............... 63

**FIFTH GRADE MATH** ............... 79

**APPENDIX** ............... 96
   *Math curriculum summary for grades one through five* ............... 97
   *A list of games for lower grades math* ............... 99
   *A step-by-step progression for the arithmetic facts* ............... 100
   *Sample drawings for fifth grade freehand geometry* ............... 101
   *Square and triangular numbers* ............... 102
   *Perfect, abundant, and deficient numbers* ............... 103
   *Powers of two table* ............... 104
   *Prime numbers up to 2000* ............... 105
   *Even numbers as the sum of two primes* ............... 106
   *Odd numbers as the difference of two squares* ............... 107
   *Numbers as the sum of two squares* ............... 108

# −Introduction−

## *The Math Tightrope*
In teaching mathematics, perhaps like no other subject, we find ourselves walking a tightrope. If we step a bit to one side, a sizable portion of the class becomes perplexed and overwhelmed. If we correct ourselves to the other side, the quicker students get bored and the class as a whole doesn't progress enough. For many teachers, each step along this tightrope brings up unpleasant memories from their own childhood.

As Waldorf teachers, we are aware that the teaching of math is more than just an intellectual exercise. We engage the will of the students as we use movement and rhythmical exercises to teach the times tables. Through appropriate stories, we engage the feeling realm of the students and spark their interest in math. Rudolf Steiner, the founder of the first Waldorf school, spoke of the importance of "permeating the soul with mathematics in the right way" and how a healthy relationship to mathematics can benefit the student's later spiritual development.

Certainly, math in a Waldorf school is viewed differently and taught differently. Yet, the problems we face are often the same as in the mainstream. All too often, a class enters middle school with many of the students weak in math and saying, "I'm bad at math, and I hate it."

Why is this so? How can we do better?

## *Who is this for?*
While this book is intended for someone currently teaching math in a lower grades Waldorf school, the material presented here can be effectively used by any teacher wishing to bring meaningful, age-appropriate math to their students.

We recognize that the readers of this book come from a variety of backgrounds – different countries, state schools and independent schools, classroom teachers and homeschool educators. It is likely that your situation is different from that of the authors, and yet we feel confident that this book is a helpful resource for all of the above situations. Creativity on the part of the teacher is always necessary for success in the classroom.

## *About This Book*
Although we have attempted to address many of the issues that we feel are important, we must acknowledge that there is much more to teaching math than what is found within the covers of a book. This book cannot be a substitute – nor can any book – for a proper teacher-training program, where much of the fundamental and necessary philosophy, pedagogical basis, and teaching technique are covered.

There is much good about what is currently happening with the teaching of mathematics in Waldorf schools. This book, perhaps, focuses more on aspects of mathematics teaching that the authors feel are often in need of improvement in Waldorf schools.

## *A Word of Caution*
The authors hope that the contents of this book will be helpful to teachers and bring to consciousness some aspects of math teaching that may otherwise not surface. We hope that this book will generate thought and discussion. This book should not be viewed as a recipe for teaching mathematics in a Waldorf school. Indeed, there is no such recipe. This book is merely a resource that reflects the opinions of the authors. There are other opinions as well that may also be called "Waldorf".

The reader (presumably a teacher) should not simply follow instructions or carry out a plan just because it is stated in this book. As always, it is the teacher's job to develop the inner sense of what is right at a particular moment for the class and to create an effective lesson. It is ultimately the teacher's love and creativity that will bring the material alive for the students. No book can give this to the teacher.

## *Resources for Teachers*
www.JamieYorkPress.com. On our website, you can view full-color photos of children's main lesson book pages, download (for free!) a variety of practice sheets, and access other resources related to *Making Math Meaningful* books. You can also contact us there with questions and comments.

Our puzzle and game book. There are times when something different is called for. Our book *Fun with Puzzles, Games, and More!* is intended as a resource for math teachers in grades four through twelve, in part to supplement the normal classroom material. It provides ideas for that "something different". Additionally, we have books for middle school and high school math. You may order these books through our website.

Online grade-specific workshops for teachers. We have online workshops for each grade. Each workshop was recorded live and then edited to create more than eight hours of video content divided into more than twenty modules. Visit our website to find out more.

Introduction

## *Acknowledgements*

The authors wish to thank the following people for their help in making this book possible: Marilyn Fox for a variety of ideas; Grace Wang, Vienna Scheyer and Katja Kramers, students of the Seattle Waldorf school for sharing their beautiful math main lesson books; Meike Gottenbos for helping with some drawings; and Ruud Luiten for technical support. And a very special thank you goes to Else Gottgens for the many wonderful discussions about child development that we had over a 35-year period.

## *What are the symptoms of our "math illness"?*

### Fear and stress

In this day and age, fear can have an influence over much of our lives, including our children's education. Teachers and parents are more worried than ever. "What will happen if my child gets behind?" The pressure to get ahead has increased. Education has become a race! More advanced material at a younger age. More homework. More high stakes testing. More fear. More stress. Even independent schools are affected.

Stress and fear are not conducive to learning anything; they impair our students' ability to learn math. We need to reduce fear and stress and create a calm, safe learning environment that enables our students to achieve an inner state of calm contemplation. This creates a vessel in which learning math is possible.

### Avoiding struggle

Ironically, even though there are greater expectations for our children to learn more material and do more homework – all at a younger age – we don't want them to struggle. We want our children to be happy and successful. We live in an "instant gratification" culture. If we want something to happen, we expect it quickly. And it should be easy. Our children should get A's, be ahead of the others, and be comfortable.

The trouble is that learning math usually isn't that quick and easy. It takes patience – patience for the student, patience for the teacher, and patience for the parents. *Working through struggle is an important part of learning math – and is a great life lesson.* One of our challenges as class teachers today is to create a safe environment – without fear and stress – in which our students can work through their struggles. This is certainly no easy task. Wouldn't it be wonderful if all of our children could learn how to calmly face their challenges and develop confidence as they successfully work through their struggles?

### Meaningless math

What is society's view of math? Math is often not seen as meaningful in and of itself. A widespread belief is that "good math" must be practical and useful. It then follows that the main reason to study math is that it can be useful for some other purpose or subject. All of these attitudes toward math can make it challenging to teach math in a meaningful way.

While we, the authors, believe that practical applications of math can be helpful, we also feel that there is a higher purpose to learning math. (See *Principles of Waldorf Education,* below.)

### The *LIST*

If people believe that math should always be practical, then it stands to reason that skill development would become the primary focus. Over time this has degenerated into the *LIST* – a horrifically long list of skills that are supposedly necessary for our children to move to the next level (e.g., middle school, high school, or college). The *LIST* manages to put pressure on teachers at all levels of education. They fear that if they don't get through the entire *LIST*, then their students won't be prepared. We are led to believe that this *LIST* is getting longer and longer, and more and more daunting. Such an over-emphasis on skill development is boring or overwhelming (or both) for many students – and for the teacher, as well.

We, the authors, believe that basic skills are important, but the list of topics needed to move forward is not horrifically long; it is actually quite manageable (see *What skills are really needed?,* below). We also believe there is much more to math than just learning skills. With so much emphasis going toward the mastery of a long list of skills, some of the more interesting aspects of mathematics become neglected.

# Introduction

## Making math procedural

The tendency in schools today is to spend far too much time on procedural skills at too young an age. Examples include borrowing in second grade, long division in third grade, and arithmetic with mixed numbers in fourth grade. Each of these topics would probably be better introduced at least a full year later than what was just mentioned. But here's the real issue: by having too much of an emphasis on procedural skills at too young an age, the children experience math as the memorization of a collection of *"blind procedures"*, likely paving the way for math anxiety or trauma. Even for those students who aren't traumatized by the blind procedures, it is likely that they will not have an understanding of what they are doing, and that later, in higher grades, they will have troubles when the mathematical concepts become more challenging.

For example, in the process of doing a borrowing problem, such as 72–58, many second or third graders lose sight of the two numbers, 72 and 58. Instead they see columns of numbers: 7 minus 5, and 2 minus 8 (which is made possible through the trick of borrowing). Perhaps worst of all, the young child then loses sight of what subtraction really is; for them subtraction has become this blind procedure called "borrowing".

Likewise, in the process of doing a long division problem, the third or fourth grader forgets that they are doing a division problem; they forget what division really is. And in the process of doing arithmetic with two mixed numbers (e.g., $8⅔ + 5½$), the fourth grader sees six numbers (8, 2, 3, 5, 1, 2) placed in specific locations. The final answer is then just the end result of a series of steps or tricks. In such cases, much of the work with fractions ends up being a collection of tricks. (See *Be aware of tricks!*, below.) The student likely forgets that the problem had anything to do with fractions and may not truly understand what a fraction is.

We believe that such procedural skills may be introduced in the lower grades, but that an over-emphasis on these skills when the children are too young often results in the children losing the big picture, not understanding the concepts behind the procedure, and losing (or not developing) their *sense of number*. All of this increases the chance of the child later becoming *math traumatized*.

## The big race

Today's society seems to picture education as a race. The names of programs like "No Child Left Behind" and "Head Start" reinforce this image. The impression we are given is that getting ahead early translates into being ahead at the finish. We disagree. Not only is there much more to a child's education than getting ahead, but pushing advanced material on young children (in first through fourth grade) does not translate into being ahead in the long run. In fact, such an approach usually results in (1) "dumbing-down" the material; (2) teaching that is superficial, uninspirational, and deadening; and (3) overwhelming many of the students. Instead, the best approach is to proceed slowly in the early grades, ensuring that the material is covered thoroughly, so that all of the students deeply absorb what is being brought to them. Then, after having built such a solid foundation, the children are set to really take off in middle school and high school. The best we can do in the lower grades, if we want our students to be strong when they graduate high school, is to give them a strong foundation in the fundamentals and, most importantly, to engender a joy for learning.

## Homework[1] and reports

Homework and reports are topics that extend further than the math classroom, but we feel that they are sufficiently important to mention here.

The trend in education today is to give more and more homework at a younger and younger age. Many teachers want to prove to the parents in their class that the children are doing advanced work. Even Waldorf teachers are caving in to this pressure. It's not uncommon today for third graders to be given homework.

We question this trend. We feel that there should be no homework until fifth grade, and, even then, it should be minimal. Any homework should help to create joy in learning. Children spend many hours in school. This is where they should learn to work hard! Children need a life outside school! At home they need to play, read, practice their musical instrument, do crafts, draw, paint, do board games[2], take part in the household chores, and spend time with friends and family. To quote Rudolf Steiner:

> *There is another thing to be considered. In the Waldorf School practically all the teaching takes place in the school itself. The burden of homework is lifted, for the children are given very little to do at home. Because all the work is done together with the teachers, the children's attitude is a quite remarkable one.*[3]

---

[1] For further reading that questions the value of homework, see Alfie Kohn's book, *The Homework Myth*.

[2] See the appendix for a list of games for lower school math.

[3] *Human Values in Education*, The Teachers' Conference in the Waldorf School, Lecture V, Arnheim, July 21, 1924.

# Introduction

It is often said that giving homework in the early grades helps to develop good work habits for later. This gives the impression that the best way to prepare children for the stressful burden of homework, which they will encounter in high school, is to have them experience this burden in the younger grades. In contrast, we feel that good homework habits can be developed in middle school, and even then, with a minimal amount of homework.

In a similar way, teachers now require the children to do written reports (e.g., animal reports in fourth grade, state reports in fifth grade, and book reports). It seems that this comes from mainstream education where there is a desire to imitate an aspect of higher-level education (i.e., research papers). It gets watered down and brought to the young children. Parents who are pushing for more academics may be pacified. But we feel that such reports have little pedagogical value and can even have a negative impact on the child's enthusiasm for learning. We have similar concerns about book reports. Of course, there is nothing wrong with encouraging children to read a book at home. But forcing children to read a book that they may not be interested in, and then burdening them with writing a report on it, is likely to dampen their enthusiasm for reading.

## Math trauma

Many adults in our society are weak in math and intimidated by it. Why does this happen? At what age did the math trauma begin? At what age does a child start saying, "I'm bad at math"? While for some students math trauma may not be apparent until adolescence, we believe that in most cases the root cause of math trauma begins between second and fourth grade.[4]

Perhaps our single greatest task when teaching math in the lower grades is to ensure that our students don't enter middle school math traumatized. But this is no easy task!

## Two ends of the spectrum

*Too much too soon.* Class teachers can be pressured by parents or colleagues to prove that their own class is strong at math, doing advanced material, and is ahead (or at least not behind) other classes. The teacher may or may not be conscious of this pressure, and much of this pressure may be self-inflicted. Because of this pressure, the teacher is then inclined to move quickly through the material and go into too much depth with material when the students are too young. All of this will likely lose a good portion of the class, leaving them math traumatized and thinking that they will never become good at math. Such a class often enters middle school with a huge disparity between those who are good at math (the "fast students") and those who aren't good at math (the "slow students"), with few students in between. (See *Avoiding the "Great Divide"*, below.)

*Math deprivation.* Don't mistake what has been said above as an argument to do as little math as possible. If a teacher avoids math, or the teacher keeps the math too simple, then the class won't progress enough, and the students' sense of number and ability to "think mathematically" will be underdeveloped. Such students will likely have difficulties in later years when they have a teacher who has higher expectations.

It's all about balance. We need to have (reasonably) high expectations of our students. Each student should be appropriately challenged in math. Developing a sense for what is right for the class is an important part of the art of teaching.

## Lack of independent, creative and flexible thinking

Certainly, one of the ultimate goals of any educational system should be that its graduates can think independently, flexibly, and creatively. Yet, we often hear from university professors and leaders in industry that today's education isn't able to adequately produce graduates who are good problem-solvers and can think creatively. Clearly, an approach to teaching math that amounts to marching through a textbook and blindly memorizing procedures is not likely to develop independent thinking.

Students are often given the impression that there is only one way to do a math problem – that math is rigid and inflexible. We need to instead look for opportunities to show that there are multiple ways to solve a problem, and elicit from our students their different ideas for doing a calculation or solving a problem. If we are successful with this in the younger grades, it can make it possible for our students to develop into independent and creative thinkers in high school and beyond. (See also *Developing flexibility in thinking,* below.)

---

[4] McLeod, DB, (1992). *Research on affect in mathematics education: A reconceptualization*. From Grouws (Ed.), *Handbook of research on mathematics teaching and learning* (pp. 575–596). New York: Macmillan.

Introduction

# *Principles of Waldorf Education*

Waldorf education's principles and unique goals are drawn from the developmental insights of Rudolf Steiner. What is listed here is an essential part of any Waldorf teacher-training program for class teachers. Some of these most important principles are as follows:

### Developmentally-based education

It is important that all Waldorf teachers (from pre-K to grade 12) have a deep understanding of child development and the development of the human being in general.

With this understanding of child development, the teacher can then become conscious of how something brought into the classroom can support the healthy development of the child. For example, in fourth grade, as the child separates from the world, the unity of their surroundings falls apart. Therefore, at this time in the curriculum, we see this reflected in the introduction to fractions, where the whole number falls into many pieces. This process with fractions helps to support the child through this transition in their life.

### Creativity and imagination

Creativity is at the core of education. Imagination is the ground from which knowledge springs. It is important for the teacher to work with enlivening images and pictures through the children's imaginations rather than using dead concepts and judgments. Through the teacher's creative work, the children engage their imagination and develop their own creative potential.

### Teaching from the whole to the part

This is one of the most important guiding principles in Waldorf education. Here are some examples of how this manifests:

- *The four processes.* By proceeding from the whole to the parts, we divide or restructure the whole into multiple parts. For example, imagine that we have 12 chestnuts and we ask the students to divide these chestnuts. There are many possibilities for dividing the whole of 12 into parts.[5] If we go from the parts to the whole (e.g., and ask, "What is 10 + 2?") we have only one possible answer. This is a materialistic gesture. If we instead go from the whole to the parts, we have a more social gesture – a "giving-away" gesture. (For more detail, see *"The Four Processes"* under both First Grade Math and Second Grade Math.)
- *The Structure of a main lesson.* It is best to start a main lesson block on the first day with a story/image which contains a picture of what the whole block will bring. This creates an imaginative picture for the students, and gives purpose and structure to the main lesson block. Then, in the following day and weeks, we construct the main lesson block, piece by piece.
- *Movement exercises.* For example, at some point in second grade, the class is learning the 6's times tables through rhythmical exercises. We start by having the whole class do the movement together with the teacher, and gradually work in smaller groups until individual students can do the movement on their own. Here are some more details…

### Rhythmical exercises

Math is naturally connected to the human body, while the language arts are connected to culture. In Waldorf schools, we bring mathematical processes to consciousness through movement.

A new visitor to a Waldorf school is impressed by watching entire classes of children doing various complicated rhythmical movement exercises. These rhythmical exercises have developmental benefits beyond how they are used to bring the mathematical concepts into the children's bodies.

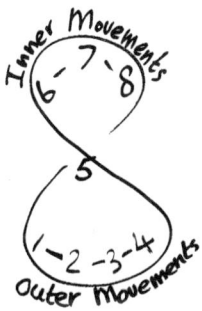

In first grade, we move math as much as possible so that the counting numbers become synchronized with the stepping and clapping of the rhythmical exercises. Through the years, these outer movements are more and more transformed into inner thinking processes. By working in this way, we hope that by middle school the children will have established more lively inner thinking processes.

---

[5] Rudolf Steiner, *Renewal of Education*, Chapter 10 (Synthesis and Analysis in Human Nature and Education), Lecture given in Basel, May 5, 1920.

## Introduction

*Progression for rhythmical exercises.* When used in the math lessons, we need to be sure that these movement exercises follow a progression that brings the children from the will into consciousness. Too often, teachers only have the whole class do some movement/speaking exercise in unison, and nothing else. This tends to put the children into a state of sleepiness. In order to bring it into consciousness, we work from the whole to the part (see above) and follow a progression similar to the following:

- The whole class does the exercise (perhaps movement combined with speaking) with the teacher leading.
- Once the children have become familiar with the exercise, they can do it together while the teacher only observes.
- Then a group of students does the exercise, while the rest observe.
- Then individual students do the exercise, taking turns.
- We close each session with everyone doing the same exercise together, one more time.

Additionally, for something like practicing the times table, after practicing moving and speaking together, we should also practice speaking it without moving. In the end, this can be further brought into consciousness with written work, practicing problems ("What is 7 times 4?"), and mental math.

### The conscious use of forgetting[6]

One might think that if we were successful in teaching the children, then they would always remember what they were taught; if they forget what they were taught, then it must be a sign that we have failed. But this is not so. *Forgetting is an important part of learning.* After teaching something new to the children, we should then let it go. It "goes to sleep" in the children and works on them subconsciously. Then, after some time, we review – bringing the old topic back and deepening it. The children will have forgotten what they had previously known, but then, during this review process, they will quickly remember it all, and have a new, deeper understanding of the material. This is all especially true when we introduce a completely new topic in a main lesson – after the main lesson is over, we let it go to sleep for at least four weeks before we start practicing the new concept.

### The pedagogical law[7]

As a central tenet of Waldorf education, the teacher educates each of the student's (finer) bodies by working out of their[8] higher body. In kindergarten and early childhood education, the young child is largely working on the development of their physical body. Therefore, the teacher works out of their etheric body while working with imitation. Prior to the teenage years, it is the teacher's astral body that aids the development of the student's etheric. After that, it is the teacher's ego that helps to develop a healthy astral body in the students.

### Educating the "freed etheric body"

During the child's first seven years, the etheric forces were used to shape the physical body. With the change of teeth, some of these forces are then freed up and available for the learning process. The education of this freed etheric in the child starts in first grade. What has been developed unconsciously in the first seven years will come to blossom in the next seven years through our work in the classroom. Beginning at the age of seven, we can begin to work with the freed etheric forces and nourish the astral body. What we are trying to cultivate is memory, lasting habits, temperaments, lasting inclinations, and enduring desires.[9] [10]

Let's be clear what we mean when we say we need to work on the child's etheric body. When most Waldorf teachers think of working with the children's etheric bodies, they tend to think mostly of *rhythm*. The teachers then tend to spend more time and energy with the rhythmical activities in the lesson than working to strengthen the memory forces.[11] Every morning in a Waldorf school, we can hear many rhythmical activities, mainly done by the teacher, where the whole class follows at will. In these cases, the children often don't work enough individually with the material. (See *progression for rhythmical exercises*, above.)

---

[6] Rudolf Steiner, *The Being of Man and His Future Evolution: Forgetting.* Lecture given in Berlin, November 2, 1908

[7] Rudolf Steiner, *The Education of the Child in the Light of Anthroposophy*, 1907. Discusses the birth of the human being's four bodies. These lectures were given eleven years before the start of the first Waldorf School.

[8] At the time of the writing of this book, gender has become a social and political issue. Linguistically, rules for the use of gender-specific pronouns are in transition. It is difficult to predict what the conventions in the future will be. In the meantime, we shall use the words he, him, she, her, and their in the generic sense.

[9] Rudolf Steiner, *The Etheric Being in the Physical Human Being.* Lecture in Berlin, April 20, 1915 GA 157.

[10] J. Bockemuehl, et.al., *Towards a Phenomenology of the Etheric World.* (pp 217-235).

[11] Rudolf Steiner, *Memory Persephone, Education and Teaching as Preventative Medicine.* Excerpts from Chapter 6. Published by the Medical Section, Dornach, Switzerland, 2006.

However, working with the child's freed etheric body actually has more to do with *memory, habits and imagination.* "Fully conscious repetition cultivates the true will's impulse."[12] Instead of just counting the beats by ones every day, the teacher should also work with the students counting the windows, the eyes, the noses, the fingers, the legs of the chairs, etc. "A more unconscious repetition cultivates feeling."[13]

### Review and practice

In Waldorf pedagogy, the word "review" is used in two ways: (1) daily review, which helps the child recall what was done during the previous day's lesson, and (2) review of material that was covered some time ago. Both are an important part of the learning process. Unfortunately, both are often skipped or done ineffectively, even in our Waldorf schools.

Regarding the first meaning of the word "review", Rudolf Steiner emphasized in his book, *The Study of Man*, the importance of how the teacher must be conscious of using the child's sleep life as part of the learning process. This is a central tenet of Waldorf pedagogy. Therefore, the teacher must re-view – bring it back into the child's imagination – the new material from the previous day. In doing this, the teacher should review the lesson from the previous day by using different examples. For instance, in first grade, we might regroup 12 children, and then the next day during the review we might regroup 12 chairs. Then we can go further by working with a new number.

The second meaning of the word "review" relates to the word "practice"; the children should practice material from the current block (when the class is in a math block), previous blocks, and previous grades. In today's culture, there seems to be less patience for review and practice. Teachers today are often hesitant to have their students practice very much because they feel that it isn't interesting for the students. In contrast, we believe that review and practice are vitally important for successful math lessons.

However, keep in mind that new material that has just been introduced, (for example) in math block #1, should be put to "sleep" until math block #2, when it can then be reviewed and practiced. Then, after the conclusion of math block #2 (and in grades 2-5, when the class is in a non-math block), this material (that was introduced in block #1) can be occasionally worked into the daily 10 minutes of math practice time.

Students can only learn math skills through adequate and systematic review and practice. The teacher needs to ensure that the students review and practice both the new material and topics covered in previous blocks. An old teacher's motto states that for a student to learn something permanently it should be reviewed the next day, the next week, the next month, and the next year. That is systematic review!

However, be sure that there is joy during math practice time. The problems should be interesting; there should always be something new to discover.

### Teaching economically

In a Waldorf school, we are able to teach fewer hours of math than what is done in the mainstream, and still our students do fine. We teach economically! This is possible for a number of reasons. First of all, the Waldorf class teacher stays with the class for several years, which enables the teacher to develop a deeper relationship with their students and to better know the students' strengths, weaknesses, and how they learn. In a Waldorf school, we are doing much more than teaching skills; we are developing capacities in our students. Starting in the early grades, we work on developing the students' observational skills, which allows them to more easily relate one situation to another, and thereby more easily solve a new math problem. These are just some of the reasons that Waldorf education stands out!

### A higher purpose

Rudolf Steiner spoke about how mathematics is a training in sense-free thinking. He also spoke about how the proper teaching of mathematics is an important part of the students' moral and spiritual development. By developing mathematical capacities in our students, we are helping to lay the foundation for the students' spiritual development later in life.

> "The student of mathematics must get rid of all arbitrary thinking and follow purely the demands of thought. In thinking in this way, the laws of the spiritual world flow into him. This regulated thinking leads to the most spiritual truths." -- Rudolf Steiner[14]

---

[12] Rudolf Steiner, *Study of Man*. Stuttgart, August 21-September 5, 1919, Chapter 4.

[13] Rudolf Steiner, *Study of Man*. Stuttgart, August 21-September 5, 1919, Chapter 4.

[14] Rudolf Steiner, *Spiritual Ground of Education,* lecture given at Oxford, August 21 1922, GA 305

Introduction

**The Waldorf class teacher**
All teachers have a tremendous responsibility to oversee the development of the children in their class. Additionally, all Waldorf teachers have a great deal of freedom in deciding what to bring into the classroom. They do not rely on textbooks; instead each child creates their own book for every main lesson.[15]

It is important that we, as Waldorf teachers, are the authors of what comes into the classroom. It needs to be clear to us why we do what we do. We shouldn't just blindly do something because it's what everyone else does. We need to feel it's right for the students in front of us. In this way, we take ownership of the curriculum.

**Faculty study**
As a faculty, we need to study Anthroposophy. The teaching of mathematics is one of the instruments that we use to aid the children in their development. Studying child development (e.g., temperaments, polarities, and the senses) gives us a deeper understanding of the children and helps inspire us in our math lessons. This is an essential part of being a Waldorf teacher.

**Imagination and math**
In today's society, imagination is associated largely with fantasy and not viewed as very "useful". It is therefore not surprising that in mainstream education, the development of the child's imagination can be considered unimportant. In contrast, Waldorf education holds that the development of the child's imagination is a central aspect of developing the whole child.

Until the age of 13 or 14, the child thinks mainly in pictures and images. If we want these younger children to work willingly, images will motivate them. If we want them to learn eagerly, then pictures will entice them to listen and remember. When we prepare our lessons, we should ask ourselves: "Have I found the proper images for what I want them to learn?"[16]

In terms of the teaching of math, imagination can be woven into the lessons, not just by telling compelling stories, but also by showing students that math is a fascinating and creative human endeavor. In this way, math can be a springboard for thinking flexibly and creatively.

> "Imagination is more important than knowledge. Knowledge is limited. Imagination encircles the world." -- Albert Einstein

## *Will our children be prepared?*

**Preparation for middle school, high school and college math**
Often, we hear parents who worry about their children's math, say: "Will our children be prepared?" This certainly is a fair question. We may then ask, "Prepared for what?" Usually the anxiety is regarding preparation for the next step – whether this next step is middle school, high school, or college. And today, this anxiety about "being prepared" seems to be with children at a younger and younger age.

It is an interesting exercise to ask a group of parents who have this question – and perhaps the anxiety as well – what they think it would look like to have their children well prepared in math for the "next step". Surprisingly, when brought in this way, parents won't usually speak of the necessity for their child to learn a long list of math skills, nor will they say how important it is that their child be ahead of other students.

We believe that there are four critical ingredients that our children need in order to be prepared for their next step in their math education – be it middle school, high school, or college.

1. *Enthusiasm for learning.* Enthusiasm and love for learning math will likely take any student a long way. No matter how good their skills are, no parent or teacher wants their child to be apathetic about learning.

2. *Study skills.* The term "study skills" is broad; it includes organization, work habits, etc. In the lower grades, we prepare and plant the seeds for the work in the higher grades by developing good habits, which includes ensuring that the students complete their work, that their work is well organized, readable, and that they show interest and take pride in their work. Along these lines, we should ask: "Have they learned how to learn, and do they have confidence in their abilities to learn new material?"

3. *Higher level thinking.* All too often in today's world, students graduate from high school and college unable to think for themselves. Their thinking, even when doing math, is largely imitative; they can do

---

[15] A *main lesson* in a Waldorf school takes place first thing in the morning for about two hours. A variety of activities might be done during this time, but the subject of the main lesson changes every four weeks or so.
[16] Else Gottgens, *Waldorf Education in Practice*.

math problems as long as they have seen something similar before. As Waldorf educators, our goal is to have students graduate high school who can think flexibly, creatively, and independently. We want our students to be able to think for themselves, think analytically, and we hope that their thinking is heart-felt and imbued with imagination. Once again, this starts in the lower grades. If we don't start here, it is unlikely that we will reach this goal in the upper grades.

4. *Basic math skills.* Yes, skills are important. If the students don't master the basic skills, they will not feel confident moving forward. However, the list of necessary skills needed for the next step isn't as daunting as we might be led to believe…

## What math skills are really needed?

Math educators often agree that there is too much of an emphasis on procedural skills in math education today. They say that we need to find more time to develop problem-solving skills (something that begins in the lower grades, but becomes more important later) and creative thinking capabilities. Yet, many teachers complain that there is never enough time for these "extras". We can begin to feel that there is an overwhelming amount of material that the students must learn, and, if they don't learn it all, they won't be ready for the next step. As we stated above, the actual list of necessary skills is relatively short. The below lists are simply intended as benchmarks in two-year intervals.

Skills needed by the end of fourth grade

- *Arithmetic facts learned "by heart".* By the end of third grade, the students should know all of the arithmetic facts (including addition, subtraction, multiplication and division) by heart. Work in fourth grade simply reinforces what they learned earlier. (See *Learning Arithmetic Facts by Heart,* below.)
- *Sense of number.* Teaching so that the children develop a "sense of number" is very important, but not something that comes easily. (See *Developing a Sense of Number,* below.)
- *What about the rest of the skills?* Our answer to this question may appear either shocking or pleasantly surprising: <u>procedural skills do not need to be mastered before fifth grade</u>. The key word here is "mastered". Of course, the students should be introduced to many of these procedural skills before fifth grade. (See also *Making math procedural* above.)

Skills needed by the end of sixth grade. (Note: many of the below-mentioned skills may have been introduced earlier, but should be mastered (after a brief review) by the end of sixth grade.)

- *The four processes.* This includes vertical arithmetic (i.e., carrying, borrowing, long multiplication, and long division), which was no doubt introduced in the early grades, but may not be fully mastered until fifth grade by most students. For more complicated problems, some students may not achieve full mastery until the end of sixth grade.
- *Fractions and decimals.* Fractions are introduced in fourth grade, but thorough practice of fraction skills only starts in fifth grade. It will take much of sixth grade before many students will be completely comfortable with fractions. Decimals (i.e., "decimal fractions") should be a fairly easy step if the groundwork with fractions has been laid adequately.
- *Measurement.* Measurement is introduced in third grade as a main lesson, reviewed and practiced in fourth grade, and then the metric system is introduced in fifth grade. The students should have mastery over simple conversion problems in both the U.S. system (e.g., How many ounces are 3 pounds?) and the metric system (e.g., How many meters is 5.8cm?). Estimating measurement is also important.
- *Estimation.* For example, estimating that $573 \times 42$ is approximately 24,000.

Skills needed by the end of eighth grade. (Note: many of the below-mentioned skills may have been introduced earlier, but should be mastered (after a brief review) by the end of eighth grade.)

- *Percents.* Percents should be briefly introduced in sixth grade, reviewed and furthered in seventh grade (but still kept quite simple!), and then really developed and solidified in eighth grade.
- *Ratios and proportions.* Ratios and proportions are an important theme in seventh and eighth grade.
- *Basic algebra.* This includes signed numbers, formulas, and basic equations (e.g., $8-x+3 = 2x-7-9x$).
- *Measurement.* This includes basic problems with area (e.g., of a circle) and volume (e.g., of a cube).
- *Dimensional analysis.* This mostly consists of conversions between the U.S. and the metric system (e.g., 5.2cm is how many inches?). Students should be able to use a conversion table to help with these calculations.

Introduction

**The major themes for math**

*Grades 1-4: Developing a sense of number.* These early grades are not about the mastery of written procedural skills (e.g., vertical arithmetic and calculations with fractions), even though these procedures were introduced in grades three and four. First through fourth grade is about developing a sense of number. Written procedural skills can be firmed up starting in fifth grade.

*Grades 5-6: Consolidating skills.* In fifth and sixth grade, we consolidate skills that were introduced in earlier grades and develop mastery with written procedural skills, such as doing arithmetic problems in vertical form (i.e., carrying, borrowing, long multiplication, and long division) and doing arithmetic with fractions. Sixth grade is also about seeing the interconnectedness in math (e.g., the relationship between fractions, decimals, percents, and division).

*Grades 7-9: Developing abstract thinking.* This is also the time to develop the students' confidence in their own thinking. Seventh grade is the "eye of the needle" – it is often then that the student's relationship to math is determined for high school and beyond.

*Grades 10-12: Developing logical, analytical, and synthetic thinking.* If all of the groundwork has been properly laid, then this is when we truly see the fruits of all the earlier hard work that has been put into the child's education. Academics, "real math", and independent thinking can all really take off at the end of high school as the student begins to find their own identity and destiny.

**Preparation for life**

Of course, there is much more to education than preparing our students for math in middle school, high school, or college. Patrick Bassett, former head of NAIS (National Association of Independent Schools), asks the question: "What skills and values are necessary for students to succeed and prosper in the 21$^{st}$ century?"[17]

In order to answer this question, Bassett refers to several sources, including: a Harvard-based study; the government's commission on education; a think tank for higher education; the academic testing industry; public opinion surveys; Tony Wagner's *Seven Survival Skills*; and Howard Gardner's book, *Five Minds for the Future*. What Bassett finds is a surprisingly high degree of congruence between these very different sources.

In the end, Bassett's own list of skills and values needed for the 21$^{st}$ century is:

1. *Character.* This includes qualities, such as self-discipline, empathy, integrity, resilience, courage, etc.
2. *Creativity.* It goes without saying that a strength of an arts-integrated curriculum (e.g., Waldorf) is developing creativity in its students. But it is also important that our education helps to develop creative thinking, as well. Later in life, this quality manifests as adaptability and developing an entrepreneurial spirit.
3. *Critical thinking.* This includes problem solving, the ability to analyze information (filtering, analysis, and synthesis), questioning what you are "told" in the media, and thinking for yourself.
4. *Communication.* This is more than just reading and writing. Bassett is very clear that the emphasis here is public speaking. Yet, in today's information world of media and computer screens, children (and adults!) are losing their abilities to speak articulately. Think of how beneficial it is to have the students, year after year, stand in front of an audience and speak – be it in a dramatic play, or at an all-school assembly.
5. *Teaming.* The ability for people to work together productively and collaboratively has become more important than ever. And today's youth loves working in groups!
6. *Leadership.* Today, perhaps more than ever, the world is in need of inspirational leaders, who have strong values, and the courage and will to "stand up for the good".

We believe that the above six skills and values speak for the strength (and need!) of Waldorf education.

---

[17] Patrick Bassett, *Demonstrations of Learning for 21$^{st}$ Century Schools*, Independent Perspective (Journal), Fall 2009.

Introduction

# *In the math classroom*

## Two types of topics: Skills and Mathematical Experiences

As teachers, it is only natural that our own education deeply influences our view of mathematics. For most of us, our formal math education worked largely like this: the teacher explained a new concept; the teacher gave us examples of solving problems with this new topic; we practiced the new problems for homework; after some time, we were tested; and, at some point later, the topic was reviewed. In spite of much work on teaching methods in mathematics, and many experiments with mathematics curriculum, this model for teaching math remains prevalent today from elementary school to graduate school. In many cases, this model for teaching math is both appropriate and effective; in many cases it is not.

This method for teaching math is completely focused on skills. It probably is most useful in grades seven through nine. In earlier years, such a skills-based approach should be used sparingly. In later years (upper high school and beyond), students should be encouraged to think more on their own and not just imitate their teacher.

As teachers, we sometimes think that the students are supposed to "learn" everything that we teach – that everything should be learned, tested, and retained. However, that's not true. It can be helpful to consider that math topics can be divided into two categories: *skills* (i.e., material that needs to be mastered) and *mathematical experiences*.

*Skills* (a topic that needs to be mastered). Here the teacher needs to create a dance between introducing, deepening, practicing, sleeping, and reviewing. The bigger the topic (e.g., fractions), the greater the number of times it needs to be put to sleep, and then later reviewed. It is quite typical to introduce a skills topic one year, but not have the students reach mastery until the next year, or the year after.

*Mathematical experiences*. With a "pure" mathematical experience, there is often no expectation that the students remember the topic or learn it as a skill. Examples of such topics include puzzle problems, "Wonder of Number" topics, number bases (8th grade), and a variety of geometry topics. We teach these topics because they stretch our students' minds, teach them to think mathematically, and engender enthusiasm and wonder for math.

And, of course, we should acknowledge that many topics can be considered a mix of skills and mathematical experiences.

## Learning the arithmetic facts[18]

Far too often, students going into sixth grade are too slow with arithmetical procedures (e.g., long division, long multiplication, etc.) simply because they don't know their arithmetic facts. This should not happen.

Knowing the arithmetic facts has nothing to do with how "smart" a student is. In reality, a lack of confidence often causes students to do poorly in math during their middle school and high school years. This lack of confidence often starts from them not knowing their multiplication facts, which makes them *think* that they must be bad at math, and in the end, it can turn into a self-fulfilling prophecy.

There is a fairly narrow window – the center of which is third grade – for learning the arithmetic facts. It is much more difficult to learn these facts by heart after fourth grade.

So how can we reach our goal of having all of our students learn their arithmetic facts by heart by the end of third grade? The real key to achieving this goal is working systematically and creatively with the children starting in first grade. Also, to help in this effort, we have prepared *arithmetic facts practice sheets* that can be used in third grade, and have also prepared practice sheets for review and increasing speed for fourth and fifth graders. These sheets can be downloaded for free from our website: www.JamieYorkPress.com.

For more details, be sure to read the following two sections of this book:

- *A Step-by-step Progression for the Arithmetic Facts* (in the appendix).
- *All about the Third Grade Arithmetic Facts Practice Sheets* (under *More Ideas for Teaching Third Grade Math*).

## Word problems

At times, word problems are held up as being more important than they really are. Some say that word problems show how math can be useful in the real world, and that word problems help to develop problem-

---

[18] Throughout this book, the term "arithmetic facts" refers to all of the basic facts that should be learned by heart, including the addition facts (up to 18=9+9), the corresponding subtraction facts, the multiplication facts up to the 12's table (of course, out of order), and the corresponding division facts.

## Introduction

solving abilities. Neither is necessarily true. In truth, many students (especially in the upper grades) quickly learn to hate word problems; they shut down as soon as they see a word problem.

There are many different types of word problems, and many levels. It is helpful to keep in mind that the analytical thinking abilities needed for true problem solving only really blossom toward the end of high school.

Our point here is not to say that word problems should be avoided until high school. Word problems should be brought into the classroom, starting in first grade, through stories and images. But we should be very careful! We need to keep in mind the pedagogical purpose of word problems. Word problems can help students see the interconnectedness of language – how English can be used to express a math problem. Word problems – if well-chosen – can show how mathematics can appear in our everyday lives. We need to carefully consider how to bring word problems to the class in a way that isn't traumatizing. Keep it simple, and make it fun!

### Mental arithmetic[19]

In our modern technological world, it may seem that the ability to work with numbers in your head is no longer necessary. We strongly disagree. Mental arithmetic is an important skill to be developed from first grade up through middle school. If practiced regularly in the classroom, mental arithmetic strengthens the students' sense of number, challenges their memory, increases their ability to focus, and develops general cognitive capacities. The use of calculators weakens the students' ability to do mental arithmetic. We therefore discourage the use of calculators before eighth grade.

Mental arithmetic can be worked on in a variety of ways, including orally with the whole class, or individually on paper (e.g., arithmetic facts practice sheets). It can be used for practicing skills (arithmetic facts in third or fourth grade, or simple fraction practice in fifth grade), or it can focus more on simply developing the ability to work with numbers in your head. As a guideline, mental arithmetic should be practiced daily for 10 minutes or less. Examples of mental arithmetic calculations are:

- *Arithmetic facts.* Before fourth grade, much of mental arithmetic may consist of practicing the basic arithmetic facts.
- *Number journeys.* For example, what is 5 plus 8, minus 3, times 6, divided by 2, divided by 5? Be careful though – weaker students may give up on number journeys.
- *"Halfway" problems.* These can be done for different grade levels. For example, in second grade, you could ask: "What is halfway between 3 and 9?" In fifth grade: "What is halfway between 423 and 428?"
- *Counting backwards.* This is very helpful in first grade. But even for fourth graders it can be helpful to ask the students to state the five numbers coming before a particularly large number (e.g., 3780).
- *Remembering large numbers.* This is good especially for fourth grade and up. Take a problem that everyone could easily do on paper, and have them do it in their head. This requires great concentration and memory. Examples: 78+28; 234+38; 612–67.
- *Subtracting from 100, 1000, etc.* This is good practice starting in third grade.
  Examples: 100–35; 100–72; 600–32; 1000–222; 500–374; 1,000,000–36.
- *Mental arithmetic "shortcuts".* In our sixth and seventh grade workbooks, there are 27 calculation shortcuts (referred to as "math tricks"). Some of the simple ones can be done in fourth and fifth grade.
  Examples: 300 x 5, 30 x 50, and 4.82 x 10. It is important, however, that there is a process through which the class was guided that brings the students to an understanding of why the trick works. (See *Be aware of tricks!* under *The Art of Teaching Math,* below.)

*A word of caution.* As with word problems, there is a tendency for some students to give up with mental arithmetic. So, the same rules apply: keep it simple, and make it fun! However, it can be best to challenge the students with a tough problem at the end.

### Developing a "Sense of Number"

Having the students develop a sense of number is an important goal for lower grades math. However, this is very challenging.

It seems that some students were born with a natural sense of number. And for other students it seems as if they will never develop a real sense of number or be able to think mathematically. Yet, often, if we allow the curriculum to soak in, and allow the student to slowly move along with their math work – all the while being

---

[19] By "mental arithmetic", we are referring to doing calculations in your head without writing down your work.

# Introduction

patient and not anxious – then one day (maybe in fourth or fifth grade) we are pleasantly surprised to see that this student has indeed been able to develop a modest sense of number and a healthy relationship to math.

Certainly, some of what we do with our math lessons (e.g., mental arithmetic) directly helps develop the children's sense of number. Yet, just as importantly, and less well understood, is what should be avoided. What follows is an incomplete list of things that can help develop the students' sense of number:

- *Do daily mental arithmetic.* Beginning in second grade, this may be the best way to directly improve the students' sense of number. (Also see *mental arithmetic* above.)
- *Number dictations.* Learning to read and write numbers up into the millions helps the child to develop a sense of number.
- *Estimating.* Starting in first grade with estimating the number of apples in a basket, or the number of people in a room, continuing with larger estimations in second grade, estimating measurement starting in third grade, and estimating the answers to difficult vertical arithmetic problems – estimating is a great way to strengthen the students' sense of number. There are endless possibilities for bringing estimation into the classroom!
- *Don't force the memorization of the arithmetic facts too early.* Often parents and teachers become nervous when they see children using their fingers to do calculations. The real question here is: at what age is counting on your fingers acceptable, and when should a student no longer be doing this? We feel that in second and third grade (and perhaps even into fourth grade for some students) using fingers to calculate can be positive; this can mean that they are thinking through the problem, which in general is good for developing a sense of number. If they were to memorize the arithmetic facts in second grade, then they would miss out on the benefits of "figuring it out for yourself" (e.g., instead of memorizing 8+5 = 13, think: "from 8, I go up by 2 to get to 10, then another 3 to get to 13"). This can do wonders for developing a sense of number!
- *Allow for different methods.* By showing different methods to do calculations, students learn to be flexible and creative in their thinking, which can then help to develop a sense of number. Of course, there needs to be balance. If we always show just one method to solve a problem, then the students can become inflexible in their thinking. At the other extreme, showing different methods to solve almost every problem can confuse and overwhelm many students. (See also *Developing Flexibility in Thinking*, below.)
- *Keep in the horizontal as much as possible.* We believe that this is very important. Far too often, students are trained at too young an age to write all arithmetic problems in vertical form. Therefore, 73–65, and 65+8 are both written vertically. Instead of *thinking* about the two given numbers and trying to figure out what the answer is – which, of course, helps to develop a sense of number – the problem just becomes a carrying or a borrowing problem, and the student is instead just working with columns of numbers. Therefore, we recommend that carrying and borrowing *not* be introduced until third grade. Even after carrying and borrowing have been introduced, it is important to have the students continue doing arithmetic both in the horizontal form (with practice sheets) and orally with mental arithmetic.
- *Minimize blind procedures.* Related to what has just been said above…in general, procedures where the children follow steps blindly inhibit the development of a sense of number. This does not mean that such procedures (e.g., borrowing, long division, arithmetic with fractions, etc.) should not be done at all. But it may be a reason to delay the introduction of such procedures, and even after a procedure has been introduced, we should be careful to not overemphasize it. We need to be sure that the students still maintain their ability to do calculations the "old way" (i.e., horizontal arithmetic and mental arithmetic). (See also *Making Math Procedural* above.)
- *When in doubt, wait*! In general, when the students are introduced to a topic before they are developmentally ready, then they will tend to simply follow the procedure without really understanding what they are doing. We should *not* be asking: "How soon can I get the children to learn this topic?", but rather: "Are the children developmentally ready to receive this topic?"
- *Allow for discovery.* There are many times when it is completely appropriate for the teacher to show the class how to do something in math and then have the children practice it. However, it is important to look for opportunities for the students to discover things on their own. By discovering things on their own, students become excited about learning and the material becomes more meaningful. It also helps to develop flexibility in their thinking and helps to develop their sense of number. Examples include:
    - With addition and multiplication (around second grade), knowing that the terms can be reversed and the result will be the same. (E.g., 6x10 is the same as 10x6.)

- When multiplying with zeroes (around fourth grade), knowing that zeroes can be ignored and then added back later. (E.g., 40x70, 400x7, and 4x700 are all equal to 2800.)
- When subtracting two large numbers (starting around second grade), we can instead add to get the answer. (E.g., with 73–68, we can think of it as taking 2 steps from 68 to 70, and then 3 steps from 70 up to 73. Our answer is then 2+3, which equals 5.)
- When dealing with cost problems (around fifth grade), we can first determine the unit cost of one item and then use this to determine the price of a certain number of that item. (E.g., if oranges are being sold at a rate of 5 for 75¢, then we first calculate that the unit cost is 15¢ per orange, so 8 oranges would cost 15x8, which is $1.20.)

## *The art of teaching math*

### The three myths of math

*Myth #1. Only people born with math ability can become good at math.* In reality, hard work can help develop math ability. In such cases, a student in the upper grades who is seen as being strong in math, might very well have once struggled with math. On the other hand, it can happen that a student, who had strong math ability but never worked at it, ends up quite weak at math.

*Myth #2. Confusion is bad.* In reality, confusion is part of learning math. Every time we learn a new topic in math, we must go through a period of confusion as we gain clarity. A key difference between a student who is strong at math versus a student who struggles is that the strong student is usually completely comfortable with working through this period of confusion as they come to clarity, whereas the struggling student often gets frustrated and shuts down. (See also *Working with Struggling Students,* below.)

*Myth #3. Forgetting is bad.* In reality, forgetting is an important part of learning. As teachers, we can sometimes think (perhaps subconsciously) that if we have taught something well, then the students will learn it and won't ever forget it. In Waldorf education, we believe it is important to work with the students sleep life. It is good and normal for the students to forget something that they have been taught. And if it is something that we want the students to learn permanently, then we need to review it. In fact, for important topics, it can be said in order to learn it permanently, you need to forget it three times!

### What does it take to be a good math teacher?

Many class teachers feel under-confident in their own math skills, and, in some cases, had traumatic experiences with math when they were in school. Often, this results in the teacher developing an antipathy towards math. However, if a teacher has had negative experiences with math in their own schooling, it can work to their advantage when teaching math. That teacher may be motivated to "do it differently" and ultimately be determined to find a way to ensure that the class won't have similar negative experiences. Additionally, if such a teacher can find a way to rise above their antipathy toward math, then that teacher may find a newfound joy in math, which can result in bringing math lessons to the children with wonderful enthusiasm.

Still, we should answer the question: "What makes a good math teacher?" We believe that the following list includes the most critical elements:

- *The teacher needs to be enthusiastic about learning math.* For many teachers, this amounts to finding a new relationship to math. How wonderful it can be to find out that math can be interesting and rewarding!
- *The teacher needs to be able to present the material effectively.* This is the art of teaching.
- *The teacher needs adequate time to prepare for the math lessons.* For many teachers, this is the most difficult issue. With everything that is demanded of the class teacher, there often isn't enough time left to prepare adequately for the math lessons.
- *The teacher must have a healthy relationship to the students.* This helps create a safe and comfortable learning environment for the students.

Of course, it should be noted that the above list could really apply to any subject; math is really no different in this regard.

### Communication with parents

The character, commitment, and attitudes of any parent group can vary widely from school to school, and even from class to class. The interests and concerns of a pioneering school are very different from that of a well-established school. For any school, clear communication is critically important. However, given that Waldorf schools tend to attract fairly critical parents, communication needs to be even better than elsewhere. Yet, it often isn't. Problems at many Waldorf schools are exacerbated by poor communication.

# Introduction

Parents need to be informed about what we are doing and why we are doing it. Beyond simple information exchange, there need to be clear lines of communication to allow parents to express concerns and to allow for discussion and conflict resolution. All of this may seem obvious, but is easier said than done.

In terms of our role as a class teacher, we need to keep parents informed about what is going on in our classroom. This is particularly important when we are doing something out of the ordinary – different than what the parents might expect. For example, in second grade, it would be best to inform the parents *why* we write our addition and subtraction problems in horizontal form, and why we don't introduce carrying and borrowing until third grade.

## Teaching the "Big Topics"

Given that our key focus in grades one through four is (1) to develop a sense of number, and (2) to have the students learn their arithmetic facts by heart, much of what is said here regarding the "big topics" (e.g., fractions, percents, algebra, etc.) is more of a concern in fifth grade and up. In short, there are two common mistakes made with teaching the "big topics":

*Too much too soon.* Often it is said that "fourth grade is fractions"; "fifth grade is decimals", etc. Teachers are then misled into thinking that this one topic should be the central theme and dominate the math lessons for the year. Take the example of fractions in fourth grade. Yes, it is important to introduce fractions in fourth grade. Many teachers do this well. But too often, much of the rest of fourth grade consists of going deeper and deeper into fractions, and then, after a while, much of the class drowns. It may not be recognizable that so many are drowning because on the surface they are "doing" many fraction problems and getting correct answers. For example, they may have been taught a method whereby they can follow the procedure for doing a problem like $7\frac{1}{4} - 2\frac{2}{3}$, but often they get "lost" in the process (even if they do all the correct steps and get the right answer) because they lose sight of what the problem actually was. The real goal with fourth grade fractions should only be to have the students develop an enthusiasm for fractions and to have them really understand the fundamental concepts of fractions. The trickier aspects of fractions can be adequately practiced in the coming years.

*Not enough follow-up and review.* Often Waldorf teachers do an excellent job introducing a topic in a very thoughtful and creative way, but then it is dropped – forever. Measurement is a classic example. The class is taught a wonderful measurement main lesson in third grade and then it isn't touched upon again until middle school. Instead, it should be reviewed and deepened (perhaps only briefly) every year. If we want the students to learn something permanently, then we need to create a "dance" between introducing, deepening, practicing, sleeping, and reviewing.

## Timing the introduction to a new topic

All too often, if a teacher were asked, "Why are you teaching this topic to your students now?" the response would be, "Because that's what everyone does!" An example is borrowing. Most teachers introduce borrowing (and have the children practice it a great deal) in second grade. Why? "Because that's when everyone does it." We should instead ask, "When is the best time to introduce borrowing?"

If a topic is introduced too early, or studied too deeply too soon, then two things are likely to happen: many students in the class will get left behind, and even the ones that can keep up will end up simply doing things mechanically without understanding what they are doing. They are then simply following *blind procedures* instead of developing strategies, developing mental arithmetic skills, and developing flexibility in thinking. The other extreme – introducing a topic too late – is also problematic. It is best to take the middle road and look for the developmentally appropriate time to bring the topic to the students. Then they will learn the material more deeply, in less time, and with wonderful enthusiasm.

Generally, we should ask the question, "Why do we do what we do, and when should we do it?" As teachers we need to be able to articulate our answer to this important question to ourselves, to our colleagues, and to our parents. Our answers need to be soundly based upon the pedagogical principles of Waldorf education, and upon the developmental stage and needs of the children.

Throughout the course of this book, we have made recommendations about when to introduce each topic, but in the end, the teacher needs to decide when is best for the class.

Introduction

## Regarding main lesson

The *main lesson* is a central aspect of Waldorf education. In general, the following are some of the key ideas to keep in mind when creating a main lesson for the day:

*The rhythm of daily activities* may be something like this (note that the times may vary):
- *Greeting the children.* This is a time for shaking the children's hands and having a (perhaps quick) check-in with each individual child before the day begins. (Time: 10 minutes)
- *Warm-up.* This can include morning verse, songs, poems/recitation, rhythmical exercises, and other exercises related to the subject of the main lesson. The entire warm-up part of the day's main lesson need only take about 25 minutes in first grade and even less time in older grades.[20]
- *Daily review.* The review component of the daily main lesson should enkindle enthusiasm for what they heard the previous day. We will then build something new on top of this review. (Time: 10 minutes)
- *New material.* Every day the students should experience at least a little something new! (Time: 15 minutes)
- *Practice.* During any math main lesson, we should include daily practice of the material from the current block, previous blocks, and previous grades. (Time: 30 minutes or more in fourth and fifth grade, less in earlier grades)
- *Main lesson bookwork.* The children should work with material from the previous day(s) and create beautiful work that they can be proud of. (Time: 20 minutes)
- *Story.* This is simply the story material for the specific grade (i.e., fables/legends in second grade, Old Testament in third grade, Norse mythology in fourth, Greek mythology in fifth). (Time: 15-20 minutes.)

*Flexibility and balance.* The amount of time that we spend on each main lesson activity (e.g., *warm-up, review and practice, bookwork*, etc.) may vary from day to day. Maybe the students need to do certain parts of a movement exercise a few more times, or sing a particular song again, or perhaps they need more practical work. When we see that the class becomes sleepy, restless, or not engaged, we have to adjust. If they look pale, then it may be because there has been too much intellectual or memory work going on. If they have very red cheeks, then it may be because there has been too much imagination, drawing, etc. Instead, we try to strike a balance – we like to see rosy cheeks!

*Breathing and movement.* It is very important to have a healthy in-breathing and out-breathing rhythm every day during main lesson. The teacher should use the main lesson material to keep the children moving throughout the morning. It may not be best to put all of the movement at the beginning of the main lesson. It is often best to "read" the students and look for the moments in the lesson when they need some movement.

*Starting a new main lesson block.* On the first day of a main lesson, we shouldn't start a block with a review of old material from previous blocks. Instead, we should start with something completely new which catches the enthusiasm and curiosity of the students about the new topic. Review of old material should be appropriately woven into the lessons in the following days.

## Main lesson books.
- *The student's creation.* In Waldorf schools, we don't give the students a textbook; they create their own book. The main lesson book is every student's self-made textbook!
- *Progression.* A main lesson book should reflect the progress of the lessons over the course of the four weeks. The new concepts should be clearly written down. A parent should be able to follow the line of thought that the teacher has led the class through.
- *No practice problems* should be written in the main lesson book! The students should have a separate practice book. The work in the practice book should also be quite neat.
- *Pride.* It is very important that the students (in all grades!) take pride in their main lesson books.
- *Beauty.* The main lesson books should be done beautifully. It is important to teach the students, starting in first grade, how to create a beautiful book. This includes borders, layout, titles, use of color, etc. While this artistic component is important, remember that the content is even more important.
- *Completion.* There is a tendency in some classes for students not to complete their books and for teachers to even leave a main lesson unfinished. This is very unfortunate! It is very important for students to complete their work. This builds good habits for years into the future. The teacher should frequently check the students' books, so the students get feedback during the block.

---

[20] Else Gottgens, *Joyful Recognition*.

Introduction

- *Corrections.* Starting from first grade, it is important that the teacher reads each student's book and ensures that the student corrects any mistakes.
- *Reference.* Starting in fourth grade, the students should be able to use their main lesson books as a reference.

## The progression for introducing a new concept

In general, there is a three-step process for introducing a new concept: (1) engage the feeling of the children; (2) incorporate the will through some kind of activity; and (3) bring clarity to thinking. This process can be viewed on a large scale, with a big concept over the course of several days, and it also can apply to smaller topics – for example with a puzzle or a question. Note that the first two steps (the feeling and the will) could perhaps both be done in the same day (in the case of a smaller topic), whereas the third step (thinking) usually comes the next day (at the earliest), after the students have "slept on it". All of this gives us more clarity on the purpose of a main lesson book – it brings into consciousness the work that has been done in the previous days.

Here's how this might manifest (on a large scale) with the second grade main lesson of place value. We could begin the main lesson by telling a story of dwarves who are mining for gems. Problems arise because they have so many gems, but they aren't able to count very big numbers accurately. Such a story, if told well, can engage the students' feeling. Over the course of a couple of days, the story presents a solution to their problems – they find a systematic way to count the gems, which is the core idea behind place value. Then we engage the will of the students by having them act out the problem – we give them a couple thousand gems or stones, and they gather groups of 10 into small bags, and every 10 bags into a small pail or bowl, etc. Lastly, we bring clarity into their thinking by reviewing what we have done in the past few days, summarizing it on the board, and finally the students create a page in their main lesson book.

On a smaller scale, we can also see how this process manifests in fourth grade with the multiplication of fractions. Here we can begin by presenting a challenge to students. We might say: "I wonder if any of you can figure out the shortcut for multiplying fractions…" If we bring this in the right way, it will capture their interest. (It doesn't always have to be a story that engages their feeling.) Then we lead them through a series of problems where they are trying to figure out what the shortcut is. They are actively engaging their will. Lastly (perhaps the next day), we see what the shortcut is, and summarize it in a main lesson book page, bringing clarity into their thinking.

## Stories and pictures

The way in which a topic is introduced is very important; this is the strength of many Waldorf teachers. With a mathematics lesson, however, the teacher needs to keep in mind that the stories, pictures, and images that we create for the child will likely be associated with this topic for a long while. For this reason, a story used to introduce a math topic ought to be shorter and less elaborate than most other stories. Most importantly, the story should help the students understand the essence of the concept being taught. Then, at the right time, the story should fade away, allowing for the mathematical concept to be discussed and practiced in its pure form without the framework of the story. This is because we don't want the children to become too dependent upon the story; they need to understand the essence of the concept. It should also be kept in mind that not all math topics need to be introduced with a story. Telling a story about miners collecting a large number of gems and needing to develop a method of counting may be quite helpful for introducing place value to second graders, but it may not be helpful to have an elaborate story to introduce long division – it may actually make long division harder for the children to learn.

Additionally, we should be conscious of our use of a visual picture when introducing a math concept. As teachers, we must be careful to ensure that topics of pure number, such as arithmetic, fractions, percents, etc., are developed in the child's mind as free as possible from physical pictures. This doesn't mean that such pictures can never be used. Fractions are more than just pizza, but later we can point out that dividing a pizza is one way to see a fraction. Using manipulatives to help learn addition can be very helpful, but the students also need to see addition done in many ways, never becoming overly reliant on any one physical representation.

If the teacher is cognizant of all of this, it can help the students to better reach the pure concept behind the particular mathematical topic.

## Introduction

**Regarding number lines and manipulatives**

The number line (which is when the counting numbers are marked and labeled along a line) may be introduced in the middle or end of first grade and then used periodically through second grade as one of several different manipulatives. Number lines, like any manipulative, should not be overused, otherwise the children could rely on that one manipulative too much, and then it could inhibit the child from getting to the pure number concept. Instead, by seeing things in many different ways, the child can then come to understand the pure concept more deeply, and, in the process, become more flexible with their thinking.

It is perhaps best for the number line to be phased out at some point in second grade. (In fact, we recommend that the number line not be used to introduce negative numbers in seventh grade.) After second grade, it could be just fine for the student to not see a number line again until the introduction to Cartesian geometry (i.e., graphing equations) in high school.

In general, manipulatives should be used rarely after second grade for the purpose of learning number concepts and basic arithmetic. Of course, manipulatives do serve quite a useful purpose with more "physical" math topics like measurement and many topics in geometry.

**Practical math**

To some degree, tying into what has just been said above, we often hear people say that math should always be "hands-on" and practical. Certainly, there are times when math can be practical, and times when a hands-on activity facilitates learning. But we shouldn't feel a need to always make it so. There is certainly lots of great math, which is neither practical nor hands-on. When planning for a lesson, we should always ask ourselves: would a hands-on activity, a manipulative, or a story be helpful? Maybe the answer is "yes", or maybe "no".

**Daily math practice**

In grades one through five, math lessons consist of an interweaving of review, rhythmical exercises, daily practice and bookwork. We recommend doing math practice every school day throughout the year (except for first grade, when math is only done during the math main lessons). As a general guideline, this math practice should take about 20 to 30 minutes per day during a math block, and 10 minutes per day during other blocks. The material for daily math practice comes from the current block (when the class is in a math block), previous blocks, and previous grades.

**Be aware of tricks!**

Too often, students are taught *blind tricks* as a means for solving a math problem while having no idea why or how the trick works. We believe that such blind tricks should generally be avoided. However, the same trick could be taught in a better way such that the students understand what is behind the trick – or, better yet, the trick may be a shortcut that the students discovered for themselves. This should be our goal; it then has become a *strategic trick* instead of a blind trick. Even if the discovery of the trick is led by a few of the "quicker" students in the class, it is valuable for the whole class to see that it comes from a process.

For example, consider this problem: Convert $4\,^3/_5$ into an improper fraction. If the students are just told that you need to multiply 4 times 5 and then add 3, then it becomes a *blind trick* because they don't have an understanding of what they are doing. However, if they have been led through the process with the teacher, and the class as a whole has come to an understanding of why it works, then this trick has become part of a strategy.

Other tricks include: borrowing, dividing with fractions (i.e., when you flip the second fraction and multiply), and cross-canceling when multiplying fractions.

Again, it's not that these tricks can't be done with the children. The point here is that rather than having the teacher introduce a trick by telling the children what the trick is, the children should feel that they played a part in the discovery of the trick. This makes the difference between a "blind trick" and a "strategic trick".

Introduction

**Developing flexibility in thinking**

Unfortunately, many students have the impression that for any given math problem there is only one correct answer and one way to derive it. Math is then seen as a rigid and unimaginative subject. We must then look for opportunities to counter this impression. This is one aspect of the wisdom of working from the whole to the part. During a first grade math lesson, if we ask, "What is 12?", think of all the possible correct answers!: 4x3, 6x2, 9+3, 15–3, etc. Then in second grade, it can be fascinating to ask the students how they can calculate 23–15. In fourth grade, the teacher can show different ways to do long division – perhaps there is a parent in the class from another country who can demonstrate how she was taught to do long division. In these ways, the students can then see that math is flexible, creative, and imaginative. This also makes math more interesting for the students, helps to develop flexibility in their thinking and, in turn, furthers their sense of number.

**Different levels, and avoiding the "Great Divide"**

We can feel pressured to have our math class move more quickly and show that "my class is indeed doing advanced math". By doing this, we may impress a few parents and colleagues, and it may make some of the quicker students happy, but we are much more likely to overwhelm some students and increase the disparity between the stronger and weaker students. This will also likely lead to more students relying upon blind procedures rather than understanding the concepts. (See *Math Trauma* above.)

By fourth grade, the level between the students usually begins to widen. Hopefully, every student is comfortable at their own level. We need to be careful that we don't increase this "divide". All too often, classes enter middle school with a "great divide" between those who are good at math and those who are traumatized by math. Usually, this extreme situation can be avoided. The key to avoiding the great divide is to make sure the "slower" students aren't getting lost and overwhelmed.

But then the question that can be asked is: "What about the quicker students? Won't they be bored?" A few things can be said about this. First of all, in the lower grades, when a student says: "Math is easy", it usually isn't really a complaint. That student is usually quite pleased with understanding the material so well. And often, when a student in second or third grade is complaining about being bored, there may be something else going on. It could have something to do with being exposed to media at home – too much fast and short visual impressions through TV or computers can make it difficult for students to think and concentrate at school. Some students are so used to being entertained all day that they cannot entertain themselves, making school seem boring in comparison.

However, the quicker students also need challenges (especially after third grade); it is important to keep these students engaged. The students shouldn't always be doing the exact same thing. At times, we should give different problems to students of different levels so that the "slower" students don't get overwhelmed and the quicker students are adequately challenged. For example, with a practice sheet that has vertical arithmetic problems like 73+45, we could add a few challenge problems, like 5736+2658, at the end. Or, from time to time, the quicker student can help out a fellow classmate. Or maybe a student needing "something extra" can even make up a math sheet for the next day.

This is all part of the art of teaching – how can we reach our goals without losing the students that tend to struggle?

**Be aware of using math in a competitive way!**

While games and competition can be fun for many of the students, they can leave a few students feeling bad. For this reason, we encourage teachers to be very discerning about what games they bring into the classroom. Simply ask yourself, "Could this game make someone feel bad?" We especially discourage games where students are required to stand up in front of the class (e.g., math-around-the-world, or math-baseball) and compete against one another. For some students, to lose in front of everyone in this way can be quite humiliating. On the other hand, a simple math board game played between friends could be fun and instructive.

In a similar way, we discourage posting in the classroom a list with names of the students who have mastered certain times tables. Imagine how the last one feels!

Additionally, before fifth grade, it is good to find ways to keep it a secret who is taking the longest to complete their math practice sheets. One way to do this is to instruct the students to immediately begin another task (e.g., reading a book) quietly at their desk so that it isn't obvious who the slowest students are. It can also be fine to only allow a certain period of time to work on a practice sheet or a set of problems; when the time is up, then the students stop working. It may be fine if some students get further along than others.

Introduction

**Regarding testing**

In mainstream education, standardized testing has become more pervasive than ever. Indeed, it drives much of what is done in the classroom. We believe that such testing is detrimental to the children and to the learning process. The major purpose of "testing" should be that the teacher is informed of the students' progress. In this sense, we should be "testing" every day (e.g., through oral exercises and practice sheets – and keep our own records of the students' progress), but the students should never be conscious of any testing. We should create an attitude in the students, beginning in the early grades, such that they naturally want to do their best, and show their teacher and themselves how much they have learned.

**What to do when things aren't going well...**

- *Relax*. Parents are often fearful. This gets through to the students and pressures the teacher, who in turn becomes fearful and anxious. Try not to have your teaching driven by anxiety and fear.
- *Parent communication*. Speak to parents, both in parent evenings and individually. Parents need to understand clearly what you are doing and why, especially if it is different than the norm. Parents need to be well informed. They need to know what their child has achieved and what their child needs to work on.
- Ask yourself: *"Why am I doing this?"* The answer shouldn't be: "because that's how everyone does it." We should have a real pedagogical answer that we can give to ourselves, our parents, and our colleagues.
- *Give it a rest*. Re-evaluate what you're doing. Doing more isn't necessarily helpful. Maybe if a particular topic is put to rest for a month or two – or even until next year – then the class will really absorb it better at that later time.
- *Make sure that the timing is right*. When should we introduce a certain topic? The question shouldn't be "How soon can I do it?", but rather, "Is this the developmentally appropriate time to bring this topic?" Live into where the children are developmentally. If it seems that many of the students aren't ready for something, then it may be wise to listen to that instinct. *When in doubt, wait!*
- *Remember your role*. Remember that one of our important roles is a coach – to help these children through their emotional math issues.

**Four steps towards math trauma**

- *Confusion. Everyone experiences confusion* when learning math. We can think of confusion as the state you are in as you are working toward clarity. Therefore, confusion is necessary as you are learning something new. Students who are deemed to be "good at math" are usually comfortable with being confused, while for other students, being confused can be bring up many emotions, which may inhibit learning.
- *Struggle. Most students struggle* with math at least occasionally. Students may struggle for a period of time until clarity is achieved – and that's not necessarily a bad thing. One of our tasks as teachers is to help our students work through their struggles so that they can experience true success.
- *Frustration. Many students get frustrated* with math. Typically, students become frustrated after struggling without success for an extended period of time. Some students get to the level of frustration almost as soon as they get confused. For many of these students, we need to support them in overcoming frustration.
- *Shutting down/trauma. Some students shut down and become math traumatized*. After a while, if a student is constantly frustrated with math and doesn't ever quite understand the concepts, then she starts to believe that it's impossible and gives up. For students in this situation, we need to develop and implement a more comprehensive plan that enables them to develop more confidence in themselves and form a more positive relationship to math.

One of the jobs of a class teacher is to guide the students through the challenges they encounter in school and in life. In today's world, our children are often shielded from discomfort; they come to expect that things should be easy. They have little patience. It is therefore not surprising that so many children have little tolerance for confusion and struggle. If we compound this with a variety of emotional issues, then we can see how important it is for us, as teachers, to *coach* these students through these challenges.

Introduction

**Perfectionism and fear of failure**

Some students have a tendency towards perfectionism. Often it is a girl who wants everything to be in order. Often, it is one of the stronger students in the class. Perfectionists want life to be rhythmical and predictable. And in terms of schoolwork, they want to do well in everything. When faced with a task, they need to be confident that they can succeed.

The core of the difficulty is that math often violates these rules. Confusion and learning to overcome your struggles are part of doing math. To some degree, especially in middle school and above, math is about facing the unknown and figuring out how to solve unfamiliar problems. To a perfectionist, all of this is threatening.

And then the student enters adolescence, becomes more self-aware, and math becomes even more unpleasant and scary. Such a student may eventually just give up on math. Why? Because it is less painful to give up than it is to put forth effort and, in the end, perhaps fail anyway. This is known as "fear of failure".

So what can we do as teachers? Once again, our role as a coach is critical. And this is bigger than just math. We need to look for situations where *they feel they can't succeed*, but *we know they can*. Such a situation could be having a role in a class play, climbing a mountain, riding a bicycle (or a unicycle!), or solving a particularly daunting math problem. They think they can't, and we know they can. At the right time, and in the right way, we need to gently push them into these situations. Our role is to coach them through their challenges, and to help ensure their success. And once they succeed, celebrate! "Look at what you did! And you thought you couldn't do it!" After that, we can build on this success. In this way, the student's confidence is slowly built up, and, over time, the student becomes more comfortable with the challenges that life presents – and hopefully becomes more open to learning math.

**Remediation**[21]

Meeting the needs of a child with learning challenges can take a good deal of time and energy. In most cases, the problems are in the realm of the lower senses – be it a problem with balance, sense of life and/or gross-motor movement. Often, these children cannot even count one step while saying "one". This makes it more difficult for them to develop a sense of number.

In general, we can always expect to have students who are somewhat behind the rest of the class. In this case, it is especially important for the teacher to know where the child is at, to understand why there is an issue, and to have a plan of how to move forward.

Confidence is an important factor. We need to figure out where the child gets stuck and pick it up from there, so the child starts to build confidence. The student needs to know that her teacher believes in her. The child shouldn't feel that she is doing "dumbed-down math". If the child (or the class as a whole) is behind or has certain holes in her knowledge, then we still have to find ways to bring the current material to her.

Dealing with all of this shouldn't just fall on the shoulders of the class teacher. There are often additional hindrances that obstruct the learning of math. As a faculty, we can work together (e.g., with a *child study*) to figure out what is behind this.

Individual *coaching* often makes a huge difference. The class teacher should ask the remedial teacher for help early on; waiting until fifth grade is too late. Any emotional damage takes significant time to heal and can further delay the child's progress with math. Some children with remedial needs (especially if we are able to work with the situation effectively) can be behind the rest of the class for years, but then they can "wake up" to math in a later grade – maybe in fifth grade, or even as late as eighth grade.

Additionally, we need to be careful not to allow fear to set in with ourselves, the student, or the parents; fear exacerbates the situation. We need to work productively with the parents of a child with remedial needs. Clear communication with parents is critical. This includes having a written record (as a school document) of the child's progress and a written record of any modifications in the child's program (i.e., an ILP).

Depending upon a school's resources, it may not be feasible to keep a child with serious learning challenges in the school. Sometimes, in spite of our own emotional attachments to the child, we have to admit that we can't meet a child's needs.

---

[21] For more ideas on working with children with learning challenges, see Ernst Schuberth's book, *Teaching Mathematics for First and Second Grades in Waldorf Schools*, pages 59 and on.

Introduction

**Seven Questions to Ask Yourself Each Day**[22] while reflecting on your last main lesson and preparing for your next main lesson:

1. Did I give an image to the students today? (Not just in the story.)
2. Did I use the night? Was everything that we did yesterday touched upon today?
3. What did I bring new today? What was new to them? What shall I bring them new tomorrow?
4. Did I translate any learning material that I possibly could, into movement? What can I do tomorrow in that respect?
5. Did I address one or two temperaments specifically? (Not only in the story!) What about tomorrow?
6. Did I see any child that did not make an effort during the day? Do I know the cause of it? What shall I do about it tomorrow?
7. Did I make the children laugh? Can I include something in my preparations that will make them laugh?

**Golden rules**
- *Healthy Math*
    - *Choose developmentally appropriate material.* Is the material meeting the children? How does this topic affect the children developmentally?
    - *When in doubt, wait!* Would the children be more able to receive this if you waited until later?
    - *Less is more!* Are you overdoing a new topic? Are you simply "plowing through" the material?
    - *Minimize blind procedures and tricks.* Are the students blindly following procedures and doing tricks without understanding? If so, could this be done differently?
- *Healthy Students*
    - *Enthusiastic students.* Are the students interested and enthused?
    - *Challenging the students.* Are you challenging the students appropriately?
    - *Avoiding trauma.* Are they getting it? Is it too much for some? Are some students' confidence and enthusiasm being damaged? Is the learning environment healthy, so that students can learn?
- *Healthy Teacher*
    - *Enthusiastic teacher.* Are you interested and enthused?
    - *Health and balance.* Are you healthy and balanced with a strong inner life?
    - *Relationship to the material.* Are you the author of what you bring to the class? Are you clear about why you are teaching this topic now to these students? Do you see the "big picture"?
    - *Relationship with the children.* The material is not the most important thing. What matters most is how you bring it, who you are in front of the children, and your relationship with the children.

---

[22] From a mentoring course by Els Gottgens, March 2005.

# –First Grade Math–

## *Overview of Child Development in First Grade*

With the change of teeth, the capacity of thinking reveals itself, and the desire to learn and to follow a beloved authority becomes apparent. Instead of imitating the surrounding world at will, the children focus more on following their teacher. Thus, the first school year is still marked by a "dreaming living" in the world. This is a ripple effect from the early childhood years. There is an eagerness for learning and the children want to engage through their feelings. They connect with the subject matter through rhymes, music, movement, etc. Through images, the spoken word enlivens the inner thinking of the children. As teachers, we want them to work willingly, so we use images to motivate them. Instead of saying: "Write your numbers straight", we say: "Let all your numbers stand as straight as a spear." Of course, we present everything with joy and enthusiasm.

For teaching math in first grade, this means we use stories (e.g., a baker who makes muffins and puts them into rows of two, or on plates in groups of four, or sells them in groups of three). The students illustrate these stories in their main lesson books. The teacher can vary things every day while working with the same concept and story.

## *Curriculum Summary for First Grade Math*

<u>The world of numbers</u>

- *Roman numerals*. We begin with the Roman numerals because they are related to the human body. The standard Arabic numerals can be introduced simultaneously with, or shortly after, the Roman numerals.
- *Quality of numbers*. What is the quality of numbers that we see in the world?
- *Counting*. The children need to become "comfortable"[23] with counting forward and backward to 100.
- *Rhythmical and skip counting*. We do rhythmical and skip counting with the 2's, 3's, 4's, 5's, and 10's.
- *Number dictations*. Number dictations should start in the second math block.

<u>Beginning calculations</u>

- *Introducing the four processes*. By the end of first grade, the students should be fluent with sums (e.g., 8+5) up to 24, and comfortable up to 100.
- *Regrouping numbers*. For example, how can we regroup the number ten? There are many answers: 2 groups of 5, 5 groups of 2, 6 and 4, 7 and 3, etc.
- *Learning the "easy" addition facts*. All addition facts up to 10, as well as all of the doubles (i.e., 6+6, 7+7, 8+8, 9+9) should be learned by heart by the end of first grade.

## *Recommended Reading for First Grade Math*

- Henning Anderson, *Active Arithmetic*. ASNA Publications.
- Herman von Baravalle, *The Teaching of Arithmetic and the Waldorf School Plan*. Waldorf School Monographs, 1967, third edition.
- Irene Groh, Mona Ruef, and the Medical Section at the Goetheanum, *Education and Teaching as Preventive Medicine*. Persephone, Forest Row 2006, pages 117-131.
- Ernst Schuberth, *Teaching Mathematics for First and Second Grades in Waldorf Schools*. Rudolf Steiner College Press.
- Rudolf Steiner, *The Spiritual Ground of Education*, chapter 5. Anthroposophic Press.
- Rudolf Steiner, *The Kingdom of Childhood*, chapter 5. Anthroposophic Press.

---

[23] Throughout this book, we use the following terms: "Fluent" means they need to have it on the tip of their tongues without any hesitation. "Comfortable with" means they are familiar with it, but there might still be some hesitation. Therefore, we have three stages in learning a skill: Introduction, "comfortable", and "fluent".

First Grade Math

## *Lessons and Topics for First Grade Math*

*"It is especially important not to go on working in a monotonous way, doing nothing but adding for six months, etc., but where possible one should take all four arithmetic rules fairly quickly one after another and then practice them...We should take all four rules at once and be careful that through practice these four rules are mastered almost at the same time."*

--Rudolf Steiner (from Stockmeyer's book)

Developing a sense of number
- The sense for numbers is awakening. The goal in first grade is for the children to come to conscious counting through movement. Else Gottgens calls this "counting along the line".[24]
- *Whole to the part*. As mentioned in the introduction, working from the whole to the part is an important aspect of our work. This is especially true in first grade. For example, instead of always asking questions like, "What is 9+3?", we can instead ask, "What is 12?" Think of the possibilities!
- The children should gradually come to see a number as an entity in itself. This is important! Can they immediately recognize a group of five on a dice, a flash of three fingers, or a group of six nuts? The children can make their own flashcards from these examples.
- *Rhythmical counting*. Rhythmical counting serves as preparation for learning the times tables in second grade. In first grade, the children learn to do rhythmical counting by 2's, 3's, 4's, 5's, and 10's. (The 11's can wait until second grade.) This work should come out of the imagination. In first grade, we start by counting by ones, but emphasize the given number. For example, with rhythmical counting by 3's, we count forwards and backwards while emphasizing every third number, perhaps by clapping and saying it louder: 1, 2, **3**, 4, 5, **6**, 7, 8, **9**, 10, 11, **12**, etc. We don't need to mention that this is the 3's times table; we simply call it rhythmical counting by 3's. Later, this transforms into *skip counting*, where they simply speak the "loud" numbers: 3, 6, 9, 12. No rhythmical counting is done in first grade between the math main lessons – it goes to sleep during those time periods.
- *Movement*. Math is related to the physical body. In order to get an awareness of mathematical processes, the children need to move their body. For many first graders, it is a challenge to synchronize their movement with their voice. Their voice is usually faster than the movement of either their hands or their feet.
- *Estimating*. Whenever presenting a problem to the students, we should first ask them to estimate. We can ask: "How many steps am I from the board?" or "How many giant steps am I from the door?" In this way, the child tries to find their position in the world, which helps to develop a spatial orientation.

Use of the number line. Number lines are often used inappropriately in the first grade classroom. Our hope here is to be clear about how and when to work with the number line.
- *In the first math block*, we don't use a number line. However, do write a row of beautiful numbers from 1 to 50 on the classroom wall. This is a seed for introducing the number line in the second math block.
- *In the second math block*, we introduce the number line as a rope with numbers written on it. The children can then beautifully write this number line into their main lesson books. Although the central theme of the block is the four processes, we don't yet use this number line for calculations. However, we can use the number line (as a rope or on the wall) in a variety of ways. For example, the teacher can point to a number and the students can say that number (or the other way around). And, once the students have practiced rhythmical counting with the 4's (beginning in block #1), we can then point out these numbers on the number line. (For more ideas, see "Activities and Movement", below.)
- *In the third math block*, once we have sufficiently worked with the four processes (beginning in block #2), we can use the number line to demonstrate calculations (e.g., 8+3=11, or 9-2=7, or 5x4=20).

Mental arithmetic. Mental arithmetic is integrated into the math lessons through the use of imaginative stories.

Scheduling. The curriculum calls for 12 weeks of math in first grade, usually one main lesson block at the end of October, one in January, and one in March. A one-week review of math can be done at the year's end.

Background reading. Be sure to reread relevant sections from the *Introduction*.

---

[24] Else Gottgens, *Waldorf Education in Practice*, page 35.

First Grade Math

# Math Main Lesson Block #1 – The Quality of Number

Content overview

- *Quality of the numbers 1-12,* with Roman and Arabic numerals. Find an image for every number, perhaps using an ongoing story where the central question is: Where is there only one of something in the world? Where is there two? Where is there three, etc.?
- *Roman numerals.* The Roman numerals naturally come out of the human body. The Roman numeral V can be seen in the hand, and X can be seen when we cross our forearms. In first grade, four should be IIII (rather than IV), and nine should be VIIII (rather than IX). The (standard) Arabic numerals can be introduced simultaneously with, or shortly after, the Roman numerals.
- *Counting.* Practice counting up to 24 (forward and backward) comfortably, and up to 100 with joy!
- *Rhythmical counting.* During this first block, we can do rhythmical counting by 2's and 4's. We should practice going both forwards and backwards.
- *Regrouping.* The idea of regrouping is briefly introduced here. It quite naturally comes out of our work with going from the whole to the part. (E.g., "How can we regroup the number ten?" There are many answers: 2 groups of 5, 5 groups of 2, 6 and 4, 7 and 3, etc.)
- *Writing.* Practice writing the numbers 1-12. Always start at the top (of the digit's written form) and draw towards the bottom.
- *The five-structure.* This is also the first visual encounter with the five-structure. The five-structure is the five fingers on our hand; two hands make ten. It is easy for most children to work with units of five. This becomes quite useful later when we start to develop strategies. It is perhaps the first strategy they will use.
- *A row of numbers.* We should put a row of beautiful numbers from 1 to 50 on the classroom wall. It is helpful to have the five-structure visible in this row of numbers by having all the numbers in the 5's (5, 10, 15, etc.) in one color, and the rest of the numbers in another color. In block #2, it can be a special surprise to notice that this coloring of the row of numbers shows rhythmical counting by 5's.

The four processes

- The four processes are briefly introduced toward the end of the first math block, but only orally (as explained below) as opposed to in a formal, intellectual manner.
- Steiner emphasizes that the very first time we introduce these mathematical processes it should be through carefully worded questions[25], as detailed below.
- These questions (for briefly introducing the four processes in block #1) can come out of the central story for the main lesson block. In block #2, the story (perhaps a completely different story) for the four processes centers around four characters – one for each process with the corresponding temperament – but this waits until block #2. (See block #2 for more details.)

*Addition*

- The very first question that we pose is important. In the first question for introducing addition, we give the sum (which is the "whole" – normally thought of as the answer to an addition problem) as part of the question, and ask what the addends (the "parts") could have been. This question can simply be woven into the central story of the main lesson.
- For example, we can tell the story (as suggested by Ernst Schuberth) about a brother and sister who just bought 8 loaves of bread and need to carry the loaves in their backpacks up the mountain to their home. We then ask: "To share the work, how many loaves of bread can each child carry in their backpack?" The question can then be answered in many different ways. Maybe the brother carries 3 loaves and the sister carries 5 (or maybe they each carry 4 loaves, etc.).
- We can then speak the statement: "8 is 3 plus 5".
- The students can then experience this as a hands-on activity, perhaps using gems or stones. Then we can ask the students about all the answers they found. Again, all of this is only done orally. We don't write it down formally (8 = 3+5) until the second math block.

---

[25] For more details, see *Discussions with Teachers,* Discussion Four, Stuttgart, August 25, 1919.

# First Grade Math

- On the second or third day, they can draw a beautiful picture in their main lesson book of the brother and sister hiking up the mountain with their backpacks.
- However, after this first question for addition has been introduced, we don't have to remain stuck only asking questions that work from the whole to the part. We can very soon (maybe after a day or two) ask questions like: "If Juan has 6 stones, and Lee has 2 stones, how many stones do they have together?"

*Subtraction.*
- Once again, the very first question that we pose is important. With subtraction, the introductory question gives the <u>remainder</u> as part of the question, and asks what must have been subtracted from the initial amount to end up with that remainder.
- For example, we could tell the following story. Emily buys 9 peaches, puts them in her bag, and walks home. After she gets home, she discovers that there are only 5 peaches left (the remainder). She didn't know the bag had a hole in it. How many peaches fell out of the bag?
- After we have figured out the answer to the question, we can make the statement:
  "5 is 9 take away 4".
- Note that this "remainder question" is different than the "normal" subtraction question, which asks something like: "What am I left with if I start with 10 gems and take away 3?" We can work with normal subtraction questions like this after a day or two – but we should first introduce subtraction with a remainder question (as given above).
- The students can then experience this as a hands-on activity, perhaps using gems or stones. But only do a few of these problems; doing too much at this point could cause overwhelm.

*Multiplication/Division.*
- Notice that we only need one question which introduces the idea behind both multiplication and division, because these two processes are deeply intertwined.
- Once again, the very first question is important. In this case, the first question for introducing multiplication/division gives two pieces of information as part of the question: the product (which is the "whole" – normally thought of as the answer to a multiplication problem) and the number of items in each group. The question then asks how many groups of this size can be made from the whole. But we need to make this approachable for the children…
- For example, we could tell the story about how Billy (who lives on a farm with lots of puppies) finds that there are 12 dog treats left in a bag. He wonders: "If I give 2 treats to each puppy, how many puppies will get treats?"
- Of course, we could also ask how many puppies get treats if we give 3 treats to each puppy? Or if we give 4 treats? The students can then experience this as a hands-on activity (perhaps using stones).
- After we find a solution to each question (given through a story problem or a hands-on activity), we can then make general statements like 12 is 4 groups of 3, and 4 groups of 3 is 12. (Note again: we only state this orally; nothing is written down.)
- Steiner suggests this kind of question for first introducing multiplication. Again, with this kind of question, we give the product and the number of items in each group as part of the question, such as: how many groups of 3 are found in 15? This question for multiplication certainly seems like a division question. This is because multiplication and division are so closely related – even more so, somehow, than addition and subtraction. This one question is a seed for both multiplication and division.

# First Grade Math

- There are three types of questions for multiplication/division. (This is only intended to help the teacher understand the finer details and shouldn't be explained to the students.)

  (1) <u>The Introductory Question</u> asks: How many groups of a certain size can be created from the total (whole)? For example, how many groups of 4 are there in 20? (Answer: 5 groups)

  (2) <u>The Standard Multiplication Question</u> asks: What is the total amount if you have a certain number of groups of a particular size? For example, how many do you have in total if there are 6 groups of 5? (Answer: 30 in total.) If it doesn't seem too overwhelming, we can ask a couple of these types of questions at the end of the first block.

  (3) <u>The Standard Division Question</u> asks: How many are found in each group if you divide the total (whole) into a certain number of equal groups? For example, how many are in each group, if you divide 20 into 4 equal groups? (Answer: 5 in each group) Compare this to the "Introductory Question", above. This type of question should likely wait until the second math block.

- *Words of Caution!* Again, we should keep things fairly "conversational" in this first experience with the four processes. We wait until the second math block to write down the formal math symbols (+, −, x, ÷) for the first time.

- The students only need to experience a few of these four processes problems (orally in story form, and as hands-on exercises) in this first block. At the start of the second block, we will recall these stories and questions from the first block.

- Language. In this first math block, it's best to use familiar language. Perhaps we don't need to mention big words like addition, subtraction, multiply, divide until the second math block. So instead of saying "3 multiplied by 5 is 15", we can say "3 groups of 5 is 15". Instead of "4 subtracted from 10 is 6" (or "10 minus 4 is 6"), we can say "10 take away 4 is 6". Then later, in the second math block, the students can become familiar with words like "subtract", "multiply", and "times", and realize that these big words mean the same thing as familiar words like "take away" and "groups of".

<u>Activities and Movement</u>

- Speaking and moving should go together. This is very important for the development of math capacities.[26]

- Count from 1-24. Practice counting by using one step per number, using images. For example, have the children imagine going over stepping-stones across a river, or dwarves stepping into a dark cave where they collect gems, jewels, etc. Have them then walk and count backwards as they go back home again.

- *Rhythmical counting*. (Hopping, clapping, stamping, etc.) Here are some examples:
  - For counting by 2's, the dwarves move the axes rhythmically: 1, **2**, 3, **4**, 5, **6**, 7, **8**, 9, **10**, etc.
  - *Even and odd numbers.* The farmer lost one wooden shoe in the heavy clay: Because the farmer (and the children) only has one shoe on, we can only hear one of the steps clearly. It goes like this: 1, **2**, 3, **4**, 5, etc., where we whisper the 1, 3, 5, and speak loudly the other ones. Of course, the children can use boots instead of wooden shoes! On the next day, the farmer might lose the other shoe/boot, so we would count the odd numbers instead.
  - *Very important!* After we have finished a movement exercise, the children should stand still and say the sequence of numbers once again, forward and backwards without movement. In this way, we bring the will activity to consciousness.
  - Read again the "Progression for Rhythmical Exercises" found in the Introduction under the *Principles of Waldorf Education*.

---

[26] Else Gottgens, *Waldorf Education in Practice*, page 37.

# First Grade Math

- *Number runner*
  - This exercise requires a line on the ground with 100 fairly evenly spaced tick marks on the line (without the labeled numbers), and with stones placed next to where the 10's would be. The first child steps and counts (1, 2, 3, etc.) on each position. Once she gets to 10, the next child starts, so that these two children step and speak their numbers – "one" and "eleven" simultaneously, and then "two" and "twelve" simultaneously, etc. When the first number runner gets to 20 (which is at the same moment that the second runner arrives at 10), a third child starts, and all three of them are stepping simultaneously, and speaking their numbers (e.g., "four" and "fourteen" and "twenty-four", etc.). It continues in this way, adding another runner every ten counts. Once a child reaches 100, he returns to the front of the line, perhaps ready to do it again when their turn comes.
- *Gems on a Circle.*
  - This exercise gives the children a geometric sense of a given number, and can be done starting with the number 3.
  - We start by having the children draw a large circle. If you are working with the number 3, you simply ask the children to place 3 gems on the edge of the circle. Naturally, they will place the gems so they are evenly spaced.
  - We can then put the triangle into movement in a variety of ways. For example, we can ask the children to very slowly move one gem until it touches one of the other two, and observe how the triangle changes shape. This brings forth an active inner imagination.
  - We can also move this whole exercise with the children standing in a circle, and three of them act as the gems.
  - After working in this imaginative way, the children can draw a triangle inside a circle in their main lesson books.
  - Rudolf Steiner writes about the importance of these activities[27]. Also, see Henning Andersen's book (p24-34) for more ideas related to this.

## Working with manipulatives

- Counting should be connected to an object, starting with their own fingers and toes. Different objects may be: chestnuts, gems, etc. Try to avoid counting with food (beans, lentils, etc.) – unless you are eating it, food is not something to play with in a world where there is hunger.
- *Regrouping numbers.* The children put both of their hands on their desk, and we ask the students to show four fingers in the air with one hand. We can then ask: how many different ways are there to show four fingers if you can use both hands? (One way would be one finger on one hand and three fingers on the other hand.)
- The teacher can also hold up four fingers, and the children need to immediately recognize that there are four fingers. This demands a lot of practice as the children learn to move away from counting. This can also be done on a table by using different objects under a cloth, etc. For examples, imagine using a certain amount of chestnuts. We show them very briefly and then cover them up. We then ask the children how many chestnuts they saw. The idea is that we want them to see the amount immediately instead of counting every single nut. This can also be done using flashcards.
- We should also ask the students to regroup a number of gems. For example, using 12 gems, there are many possibilities:

  *** *** *** ***   or   ** ** ** ** ** **   or   **** **** ****   or   ***** ** * ****

---

[27] Rudolf Steiner, *The Child's Changing Consciousness.* Lecture given on April 17, 1923. GA 306.

# First Grade Math

<u>Review and practice</u>
- *The importance of review.* We must emphasize here, once again, the importance of how the teacher must be conscious of using the child's sleep life as part of the learning process. The teacher must review (i.e., bring back into the child's imagination) the new material from the previous day, perhaps by using different materials. We might regroup 12 children, and then the next day, during the review, we might regroup 12 chairs.
- *Writing numbers.* We should practice writing numbers in a sand tray, on paper, using small blackboards, or go outside and practice writing numbers with sticks. Remember to always have the children write the numbers from top to bottom!
- We expect that simultaneously speaking and moving will be quite difficult for some of the children. Therefore, we should practice movement exercises with counting every day, but each day we use different variations. For example, we can imagine 12 dwarves going into a cave, while the rest of the students are fireflies watching them. In this way, some students are observing what the others are doing. This brings a different consciousness to the process.
- *Regrouping.* After regrouping 12 has been reviewed, we practice regrouping with a different number.

<u>Bookwork</u>
- In this block, the children can make drawings in their books from the qualities of the first 12 numbers. They can write the twelve Roman numerals along with the twelve Arabic numerals.
- The children's main lesson book should include a few examples of number regrouping.
- Be sure to visit our website (www.JamieYorkPress.com). Under the "Resources" tab you can see full-color pages from students' main lesson books.

## Lesson Plan Outline for Block #1 – Quality of Numbers

<u>Week #1</u>
- Create a beautiful story for the block.
- For this week, cover the numbers 1-4.
- Create a main lesson page for each of these four numbers.
- Roman numerals and (standard) Arabic numerals may be introduced at the same time.
- Step and count the numbers 1-24 and back.

<u>Week #2</u>
- For this week, cover the numbers 5-8.
- Continue the story for the numbers 5-8.
- Put a row of beautiful numbers from 1 to 50 on the classroom wall.
- Count many things in the classroom.
- Continue stepping and counting exercises.

<u>Week #3</u>
- For this week, cover the numbers 9-12.
- Continue the story for the numbers 9-12.
- Lots of practice writing numbers.
- Begin rhythmical counting with 2's and 4's, and other movement exercises.
- Start regrouping numbers using manipulatives.

<u>Week #4</u>
- Introduce each of the four processes through the block #1 story. (Not yet with the characters for the four processes with temperaments.)
- More practice writing numbers.
- More rhythmical counting with 2's and 4's, and other movement exercises.
- More regrouping numbers using manipulatives.

First Grade Math

# Math Main Lesson Block #2 – The Four Processes

Content overview
- *The four processes*. In the first math main lesson, the four processes were briefly introduced orally, but they weren't written down. Now, for the first time, the four processes are written down.
- *Counting*. The children should now be able to count comfortably up to 100.
- *Rhythmical counting and skip counting*. During this second block, we start by reviewing the rhythmical counting done in the first block with the 2's and 4's, and then we add the 3's, 5's, and 10's. Remember to practice going both forwards and backwards. Gradually, the rhythmical counting transitions into skip counting (see below for more details). How far we get with this depends upon the class.
- *Regrouping*. We should continue our work started in the first main lesson with regrouping numbers.
- *Writing*. We continue what we started during the first math block. Remember to always write the numbers from the top (of the digit's written form) progressing to the bottom. We can now start doing some number dictations up to 24.

New material: The Four Processes.
- At the end of the first math main lesson block, we briefly introduced the four processes through stories and hands-on activities – but it was only done orally, and reflected in their main lesson book drawings. We said things like: "7 is 9 take away 2", and "5 groups of 2 is 10". Now, in this second math main lesson block, we introduce the signs of the four processes (+, –, x, ÷), and write down the above statements as: "7=9-2" and "5x2=10".
- As we stated above in the first math block, each process leads to different types of questions.
    - *Plus/Addition*. We can ask a question like this: "How can we break 10 into two groups?", or we can ask a "normal" addition question, such as: "What is 6 plus 7?"
    - *Minus/Subtraction*. The language with minus/subtraction is particularly tricky. The teacher should use the terms "subtract", "difference", "remainder", and "minus" carefully and consciously. Of all these words, "minus" seems the most fluid. For example, think of all the ways we can think of the question "What is 11 minus 8?". We could instead say:
    "What do we take away from 11 in order to be left with 8?" or
    "What do we subtract from 11 to get a remainder of 8?" or
    "What is the remainder if I subtract 8 from 11?" or
    "What is the difference between 8 and 11?"
    All of this would be too much for first grade, but by the end of second grade the students should be able to understand all these different ways of viewing minus/subtraction.
    - *Times/Multiply and Divide*. We can ask three different questions related to the fact 3x5=15:
    "How much is 3 groups of 5?"
    "How many groups of 5 do I have in 15?" (Steiner says this is the introductory question.)
    "How many are in each group if I divide 15 into 3 (equal) groups?"
    The teacher needs to be aware (for later grades) that 20÷4 can either be asking us to divide 20 into 4 equal groups, or it could be asking us how many groups of 4 fit into 20.
- Steiner relates each of the four processes to a temperament. This can manifest in our lessons in different ways during this second math main lesson block. One way is through a story with four characters (not gnomes!) – each character representing one of the four processes and having one of the four temperaments.[28] The character for addition (plus) is phlegmatic, keeping as much as possible; subtraction (minus) is melancholic, concerned about losing things; multiplication (times) is sanguine, working as quickly as possible; and division (divide) is choleric, sharing everything fairly.

---

[28] Ernst Schuberth, *Teaching Mathematics for First and Second Grades in Waldorf Schools*. page 34-48. Also, Dorothy Harrer, *Math Lessons For Elementary Grades*. AWSNA Publications.

# First Grade Math

- *What's our Goal?* A word of caution – such a story can capture the interest of the students, but we need to eventually let go of the story, as well as any reliance on manipulatives. Our goal is for all the students, by the end of first grade, to be able to answer a question like, "what is 7 minus 3?" without picturing the character for subtraction (e.g., Minnie Minus) or counting gems. Over time, they even become able to do these simple calculations in their head, without relying on counting with their fingers.

Activities and Movement

- During this block, we play with the four processes in many ways, using manipulatives, movement, and stories. In each case, we conclude by saying something like: "And this shows that 9 minus 6 is 3." Lastly, we bring it to consciousness by writing 9-6=3.
- *Movement Activities.* Here are some ideas involving taking steps in a gymnasium or outside in a field:
    - Mark a line (with a rope) in a field. Each group of students has several stones and starts at a different place along the rope. Stepping perpendicularly away from the rope, they place a stone at every step, perhaps a total of 20 steps/stones. Then we can ask the question: "If you first take 5 steps from the rope, and then take 6 more steps, how many steps in total are you from the rope." The children solve this problem in the following manner. First, they take 5 steps, and then they take 6 steps and place a ball next to the stone where they ended up. Then they go back to the starting point and count that there are 11 steps from the rope to the ball. We conclude by saying: "Now we know that 5 plus 6 is 11", and then (after coming back into the classroom) we can write 5+6=11.
    - Another idea is to have one student (George) stand next to the stone that is 10 steps away from the rope, and have another student (Isa) stand 7 steps away from the rope. Then we can ask: "How many steps is it from Isa to George?", and conclude: "7 plus 3 is 10", or maybe some student will see how this also shows that 10 minus 3 is 7. And then we can see how this changes if Isa moves a step or two toward or away from George.
    - There are countless movement activities you can create to practice the four processes!
- *Story problems.* Give the students puzzles to solve that relate to the characters in our story, such as:
    - Peter Plus collected 5 apples in the morning and 4 in the afternoon, how many did he collect in total? Then we can say "5 plus 4 is 9", and write 5+4=9.
    - If Peter collected 20 apples today, how many did he collect in the morning, and how many in the afternoon? Think of all the answers the students might discover! To make it more challenging, we say that he collected 20 apples in the morning, afternoon, and evening (e.g., 20=6+2+12).
    - Marie Multiply has 6 hens. Today she collected 3 eggs from each of her hens. How many eggs in total did she collect? Then we can say "6 groups of 3 is 18", and write 6x3=18.
    - All of Marie's hens lay the same number of eggs each day. If she collects a total of 18 eggs each day, then how many hens does she have, and how many eggs does each hen lay each day? This is quite the good challenge for first graders. Think of all the answers the students might discover!
    - 12 rabbits were playing in the field and suddenly the fox appeared. The rabbits ran to their homes, but there were only 3 homes, and every home must have the same number of rabbits. It can also be fun for the students to act this out! At the end, we bring this to consciousness by writing:
    12=4+4+4   or   12=3x4   or   12÷3=4   or   12–4–4–4=0
- In the first math main lesson block, we did a counting movement exercise using the image of a farmer with one wooden shoe. Now we can create new stories that call for different activities and counting movements.
- We can start on a different number (e.g., 12 or 15), and count up from there until 36 or 42. We can also do this backwards (e.g., 29, 30, 31, 32, 33, 34, 35, 36, 36, 35, 34, 33, 32, 31, 30, 29). This helps the students to become confident with moving up and down past the tens (i.e., past 20, 30, 40, etc.).

# First Grade Math

- *Call and Response Counting*
  - We need to continually create new ways to practice counting. Here's one that demands a higher degree of concentration. (It might take a few days for the class to get the hang of this.) The teacher begins the counting, starting perhaps with the number 34, saying: "34, 35". The class then responds by continuing the counting, but they are required to say the same number of numbers: "36, 37". The teacher decides to continue with the next 5 numbers: "38, 39, 40, 41, 42", so that the class needs to say "43, 44, 45, 46, 47".
  - Of course, there are many ways to vary this activity. For example, the teacher could suddenly start counting backwards at any point. For instance, continuing with the above example, the teacher might say: "48, 49, 48, 47", in which case the class would reply: "46, 45, 44, 43".
  - In second grade, we can do the same exercise with the times tables.
- *Rhythmical hopping.* Another way to move the 2's, 3's or 4's is to walk and hop rhythmically along a number line (or around a circle). For example, for the 3's, we do: step, step, **hop**, step, step, **hop**, as we say 1, 2, **3**, 4, 5, **6**, etc. In second grade, we can do this with two times tables simultaneously.
- *Rhythmical counting and skip counting.* We start the block by practicing the rhythmical counting from the first block with the 2's and 4's, and then we introduce the 3's. Slowly, the rhythmical counting (with 2's, 3's, and 4's) transforms into skip counting. For example, rhythmical counting with the 3's emphasizes the underlined numbers: 1, 2, **3**, 4, 5, **6**, 7, 8, **9**, 10, 11, **12**, etc. As the days progress with this practice, the non-underlined numbers become quieter and quieter until they are just thought silently. Then it becomes *skip counting*, where we simply speak the underlined numbers: 3, 6, 9, 12, etc. With the 5's and 10's, the students can immediately jump to skip counting. Again, how far the class gets in this second math block with skip counting the 2's, 3's 4's, 5's, and 10's will depend upon the class. Of course, this work will continue in the third math block and throughout next year in second grade.
- *Rhythmical movement in a circle*
  - For the 2's. Several children stand in a circle, holding hands. With their arms swinging together backwards and forwards, they all count together – quietly saying the odd numbers when they swing back, and more loudly speaking the even numbers when they swing their arms forward.
  - For the 3's. The children are once again in a circle (this time not holding hands). Their forearms cross as their hands clap their chests and they (quietly) speak "1". On "2", they clap their thighs. Then they loudly say "3" as they clap hands with the neighbors on both sides. The rhythm continues – clapping each time they (loudly) say a number in the 3's.
- *Rhythmical/skip counting inside a circle*
  - Here is another way to experience rhythmical/skip counting. We'll use the 3's, which is represented by the circle shown here. It may be helpful in the beginning to draw a circle on the ground, but as the children become more proficient, drawing the circle may be unnecessary – the children can just see where the circle is in their imagination. Mark three places on the circle which are evenly spaced. The letters shown here should not be marked for the children (they are only shown here for the purpose of this explanation). We tell the children where "home" is on the circle (shown here as point "C").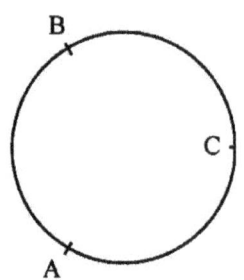
  - One child (the "number runner") starts on the circle at "home" (point C). The teacher starts counting very slowly, beginning with the word "begin": ("Begin, 1, 2, 3, 4, 5..."). The children then experience skip counting the 3's through this movement, as the number runner returns home on 3, 6, 9, 12, etc. The whole class should say the number every time the number runner passes home. So, we hear the class speaking the numbers in the 3's, and the teacher saying the rest of the numbers. Perhaps, we can even have a three-circle (perhaps not drawn?) for every child in the class so that everyone can move it together.
  - On another day, we can practice with the 2's, the 4's, or the 5's. The children will discover that with the 2's, they need to run around the circle quite quickly, and with the 4's or 5's they will move more slowly.
  - At the end of second grade, this exercise can be done with two times tables simultaneously.

# First Grade Math

<u>Working with manipulatives</u>
- Work with regrouping the numbers 1-24, using manipulatives and/or fingers.
- Students can work in pairs. For example, with the number 8, the students find as many regroupings as possible, such as: 8=2+6; 8 = 1+7; 8 = 7+1; 8 = 4+4, etc. Then the teacher writes all the different possibilities on the board (probably not in order).
- The next day, we review the regrouping that was done yesterday, and then use a T-table (as shown on the right) to organize it:
- Then we can practice these regroupings in a different way. Using your right hand, cover the section on the right side of the vertical line (we have to show this to the students) and now we will say: "8 is 1 plus what?" (they need to figure out the 7 and they can check it under their hand). Then they move down and say: "8 is 2 plus what?"

$$\begin{array}{c|c} \multicolumn{2}{c}{8} \\ \hline 0 & 8 \\ 1 & 7 \\ 2 & 6 \\ 3 & 5 \end{array} \text{ etc.}$$

<u>Review and practice</u>
- As always, the new material should be reviewed every day.
- Continue practicing writing numbers, and regularly doing number dictations up to 24.
- We should use T-tables (see above) to work with regrouping many different numbers.

<u>Bookwork</u>
- The main lesson book needs to be artistic – much more than just dry numbers.
- Have the children show the use of T-tables to regroup the numbers 1-24 (see above).
- They can begin to include some problems in their books that use the signs (+, –, x, ÷) of the four processes, such as 8=4+4, 8=10–2, 8=2x4, 8÷2=4 etc.
- Have the children draw pictures from the stories in their math lessons, which should include the characters of these stories (e.g., from the fox and rabbit story, Mini Minus, David Divide, etc.). They should also carefully draw the signs for the four processes (+, –, x, ÷).
- Be sure to visit our website (www.JamieYorkPress.com). Under the "Resources" tab you can see full-color pages from students' main lesson books.

## Lesson Plan Outline for Block #2 – The Four Processes

Week #1
- Introduce the characters for adding and subtracting.
- Play with manipulatives for adding and subtracting.
- Practice writing numbers.
- Review rhythmical counting for the 2's and 4's, and begin rhythmical counting with the 3's.
- Other movement exercises.

Week #2
- Introduce the characters for times/mult.
- Introduce the symbols +, –, x, =.
- Practice problems for adding, subtracting, and multiplying using manipulatives and stories.
- Draw the problems with manipulatives in the main lesson book.
- Rhythmical counting for the 2's, 3's, 4's, and other movement exercises.

Week #3
- Introduce the character for divide, and the symbol ÷.
- Practice problems for all four processes using manipulatives and stories.
- Write equations in their main lesson books, (e.g., 8=5 +3; 5+3=8; 15=5x3; 5x3=15).
- Practice regrouping using T-tables.
- For the 2's, 3's, and 4's, begin to transition from rhythmical counting to skip counting.
- Introduce the 5's with skip counting.
- Other movement exercises.

Week #4
- Wrap up the four processes.
- Practice problems for all four processes using manipulatives and stories.
- Continue writing equations in main lesson book
- Introduce the 10's with skip counting.
- For the 2's, 3's, and 4's, continue to transition from rhythmical counting to skip counting.
- Other movement exercises.

First Grade Math

# Math Main Lesson Block #3 – Bringing it All Together!

Content overview

- *Practicing counting.* All students should be very fluent counting (forward and backwards) up to 24, and comfortable with counting (forward and backwards) up to 100.
- *The four processes.* All students are becoming fluent with working with the four processes up to 10, and becoming comfortable with the four processes up to 24.
- *Strategies.* Introduce some strategies. This is a preview of a theme carried into second grade. In first grade, these strategies come out of our work with regrouping. For example, with 7+5, we can break the 5 into 3 and 2, and first add 3 to 7 (to get 10) and then add 2 to get a final answer of 12.
- *Rhythmical counting and skip counting.* This important work continues what was begun in the first two math blocks. By the end of first grade, it is good for the children to have worked regularly with rhythmical counting and skip counting with the 2's, 3's, 4's, 5's, and 10's.

New material

- *Progressing from counting to math.* The children now begin to move beyond counting and learn to think more "mathematically", as they develop more of a sense of number. In this way, they begin to learn strategies for doing calculations, especially involving addition and subtraction. In order to do this, the teacher should be aware of the step-by-step progression for teaching addition, subtraction, etc. (See *A step-by-step progression for addition and subtraction* under *More Ideas for Teaching,* below.)
- *Learning the "easy" addition facts.* This requires the use of memory forces. All addition facts up to 10, as well as all of the doubles (i.e., 6+6, 7+7, 8+8, 9+9) should be known by heart by the end of first grade.
- *Strategies for larger facts.* The students should begin to use strategies. For example, with 5+6 we could do 5+5+1 or 6+6–1. Using 12–4 as another example, they could think of it as 12–2–2, or even arrive at the answer by counting by 4's. Since both 4 and 12 are in the 4's table, the answer, 8, is also in the 4's table. It is wonderful for the students to share with the class, the strategies that they have discovered!

Activities and Movement

- Previously the whole class did rhythmical counting exercises together. Now we need to do this in small groups, and individually too.
- *The four steps for movement exercises.* Here is an example using 12 + __ = 19.

    Step #1:  Have the students stand on 12. How many steps are there to get to 19? In this case they are doing three things: stepping, speaking the numbers (starting with 13), and counting the steps on their fingers. In the end (after reaching 19), we can ask different questions, like: "What is the difference between 12 and 19?", or "12 plus what is 19?", or "What is 19 minus 12?"

    Step#2:  The children stand still and speak out loud the same math problem, speaking the numbers 13, 14, 15, 16, 17, 18, 19 while counting on their fingers each step – all without moving.

    Step#3:  The children sit down and write what just was spoken (12 + 7 = 19), either using paper or using small blackboards. They can also represent it on a number line.

    Step#4:  The children read back what is written down.

- *Rhythmical counting.* We continue practicing rhythmical counting with the 2's, 3's, 4's, 5's, and 10's. (See *rhythmical counting* above in block #2.) We can now work more in depth. For example, with the 2's, the whole class moves the 2's (2, 4, 6, 8, etc.), walking forwards and backwards. Then the whole class stands still while speaking the 2's. Then they sit down and write from memory the 2's on their little blackboards. Lastly, they read aloud the 2's that they have just written.

# First Grade Math

- *Games*. Try to find engaging games, including games that use the number line (see Henning Anderson's book). These can be wonderful activities for the children to do when there is extra time. Also see the appendix for a list of games for lower school math.

  *Spotting Tens*. Here is a card game for grades 1 and 2.
    - Materials: Use a normal deck of cards where the face cards and the 10's have been removed. The aces will count as a 1. Alternatively, you can use a deck of Krypto cards, or have the students make their own cards.
    - How to play: Each student has a partner – one is the dealer, and the other is the player. The dealer puts seven cards face down on the table – one card in the center, and six cards around the center card. The objective is for the player to spot pairs of cards that add to 10. When the player spots a pair adding to 10, she removes that pair, and then the dealer replaces those two spots with new cards. After the deck is finished, the dealer and player switch roles. To make the game more challenging, you can look for combinations of cards (perhaps even 3 or 4 cards) that add to a larger sum, like 12 or 15.

## Working with manipulatives

- *Phasing out manipulatives*. At this point in the first grade year, it is important to get away from working with manipulatives. The students can now imagine the problem and then find ways to solve it.

- *A string of beads*. With the smaller numbers (up to 20), we try to not use manipulatives at this point. However, from our work in the second math block, we may have identified some children who have not yet adequately developed a sense of number, and still need some visual help. Therefore, we can provide them with a string of 20 wooden beads, colored in the five-structure, as shown with the photo here.

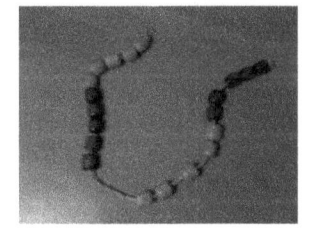

- *Number line*. This can be the right time to use the number line to demonstrate calculations. For example, we can show how 7+4=11, or 11-9=2, or 3x6=18. Also, see Henning Andersen's book (p130-134) for more ideas related to working with the number line in this way.

## Review and practice

- *Number dictations*. We can now do number dictations up to 100.
- *The four processes*. Continue writing and practicing problems that use the four processes.
- *Patterns with numbers*.
    - The children should visualize problems they did with manipulatives in previous blocks.
    - We should continue practicing the regrouping of numbers.
    - Look for ways to show the pattern of numbers. Baravalle's book has many wonderful ideas for this. Here are a couple of examples (using triangular and square numbers – but we don't need to mention these terms to the children):

        ```
        1         = 1   →              1
        1+2       = 3   →  3–1   = 2
        1+2+3     = 6   →  6–3   = 3
        1+2+3+4   = 10  →  10–6  = 4
        …etc.
        ```

        or
        ```
        1         = 1   →  1x1=1   →            1
        1+3       = 4   →  2x2=4   →  4–1   = 3
        1+3+5     = 9   →  3x3=9   →  9–4   = 5
        1+3+5+7   = 16  →  4x4=16  →  16–9  = 7
        …etc.
        ```

# First Grade Math

Bookwork
- Baravalle's book can be a nice guide for the main lesson bookwork.
- Work done with the number line can be put into the students' book in a visual way (e.g., showing rabbits hopping on the line every four numbers (4, 8, 12, etc.)).
- The children can write all of the numbers from 1-100 in a list into their main lesson books.
- The layout and use of color in the main lesson book is all very important.
- Be sure to visit our website (www.JamieYorkPress.com). Under the "Resources" tab you can see full-color pages from students' main lesson books.

## Lesson Plan Outline for Block #3 – Bringing it All Together!

To be practiced throughout the whole block:
- Rhymical counting (as needed) and skip counting with the 2's, 3's, 4's, 5's, and 10's.
- Practice writing numbers.
- Practice regrouping.
- Practice number dictations.
- Other movement exercises.

Week #1
- Start working with the number line to demonstrate calculations.

Week #2
- Continue working with the number line.
- Start working with strategies.
- Learn these addition facts by heart:
  The doubles: 1+1; 2+2; 3+3…9+9.
  Sums to ten: 1+9, 2+8, 3+7, etc.

Week #3
- Fade out manipulatives.
- Work with the difference between two numbers.
- Continue working with strategies.

Week #4
- Wrap up the block.
- Work more with strategies so that the students become more confident.

First Grade Math

# A step-by-step progression for addition in first grade

**Notes:**
- Keep in mind that these steps for addition are happening concurrently with a similar step-by-step progression for subtraction (see next page).
- This is intended to give the teacher some possible ideas of what could be done to work in a structured way.
- There are more steps possible than are listed below.

**Step #1:** Regrouping 10.
　Ask: "What is 10?"
　$10 = 1 + 9$
　$10 = 2 + 8$
　$10 = 3 + 7$
　…etc.

**Step #2:** Adding to 10.
　$1 + 9 =$ __
　$2 + 8 =$ __
　$3 + 7 =$ __
　…etc.

**Step #3:** Finding the missing number using 10.
　$10 = 1 +$ __
　$10 = 2 +$ __
　$10 =$ __ $+ 7$
　$10 =$ __ $+ 9$
　…etc.

**Step #4:** Regrouping 20.
　Ask: "What is 20?"
　$20 = 11 + 9$
　$20 = 12 + 8$
　$20 = 13 + 7$
　…etc.

**Step #5:** Adding to 20, such as:
　$11 + 9 =$ __
　$12 + 8 =$ __
　$13 + 7 =$ __
　…etc.

**Step #6:** Finding the missing number using 20.
　$20 = 11 +$ __
　$20 =$ __ $+ 8$
　…etc.

**Step #7:** Working with doubles.
　$2 =$ __ $+$ __ → $1 + 1 =$ __
　$4 =$ __ $+$ __ → $2 + 2 =$ __
　…etc. up to $20 = 10 + 10$

**Step #8:** Add two one-digit numbers for a sum between 11 and 18, such as:
　$11 = 5 +$ __ → $5 + 6 =$ __
　$15 = 7 +$ __ → $7 + 8 =$ __
　…etc.

**Step #9:** Add a two-digit number to a one-digit number, up to a sum of 20, such as:
　$10 + 2 =$ __ → $12 = 10 +$ __
　$12 + 4 =$ __ → $16 = 12 +$ __
　$8 + 12 =$ __ → $20 = 8 +$ __
　…etc.

**Step #10:** Add a two-digit and a one-digit number, for a sum between 21 and 24, such as:
　$15 + 6 =$ __ → $21 = 15 +$ __
　$18 + 5 =$ __ → $23 = 18 +$ __
　$17 + 7 =$ __ → $24 = 17 +$ __
　…etc.

First Grade Math

# A step-by-step progression for subtraction in first grade

**Notes:**
- Keep in mind that these steps for subtraction are happening concurrently with a similar step-by-step progression for addition (see previous page).
- Here again, there are many more steps and variations possible. We would like to encourage teachers to find more steps!!!

**Step #1:** The first steps.
$1 = 1 - \underline{\phantom{0}} \rightarrow 1 - 0 = \underline{\phantom{0}}$
$1 = 2 - \underline{\phantom{0}} \rightarrow 2 - \underline{\phantom{0}} = 1$
$1 = 3 - \underline{\phantom{0}}$ etc.
$2 = 2 - \underline{\phantom{0}} \rightarrow 2 - 0 = \underline{\phantom{0}}$
$2 = 3 - \underline{\phantom{0}} \rightarrow 3 - \underline{\phantom{0}} = 2$
$2 = 4 - \underline{\phantom{0}}$ etc.
…etc.

**Step #2:** Subtracting with 10 and under.
$10 = 10 - \underline{\phantom{0}} \rightarrow 10 - 0 = 1\underline{\phantom{0}}$
$9 = 10 - \underline{\phantom{0}} \rightarrow 10 - \underline{\phantom{0}} = 9$
$8 = 10 - \underline{\phantom{0}}$ etc.
$9 = 9 - \underline{\phantom{0}} \rightarrow 9 - \underline{\phantom{0}} = 9$
$8 = 9 - \underline{\phantom{0}} \rightarrow 9 - 1 = \underline{\phantom{0}}$
$7 = 9 - \underline{\phantom{0}}$ etc.
$8 = 8 - \underline{\phantom{0}} \rightarrow \underline{\phantom{0}} - 0 = 8$
$7 = \underline{\phantom{0}} - 1 \rightarrow 8 - 1 = \underline{\phantom{0}}$
$6 = 8 - \underline{\phantom{0}}$ etc.
…etc.

**Step #3:** Subtracting numbers 20 and under with a result of 10, or greater.
$20 = 20 - \underline{\phantom{0}} \rightarrow 20 - 0 = \underline{\phantom{0}}$
$19 = 20 - \underline{\phantom{0}} \rightarrow 20 - \underline{\phantom{0}} = 19$
$18 = 20 - \underline{\phantom{0}}$ etc.
$19 = 19 - \underline{\phantom{0}} \rightarrow \underline{\phantom{0}} - 0 = 19$
$18 = 19 - \underline{\phantom{0}} \rightarrow 19 - \underline{\phantom{0}} = 18$
$17 = 19 - \underline{\phantom{0}}$ etc.
$18 = 18 - \underline{\phantom{0}} \rightarrow 18 - 0 = \underline{\phantom{0}}$
$17 = 18 - \underline{\phantom{0}} \rightarrow 18 - \underline{\phantom{0}} = 17$
$16 = 18 - \underline{\phantom{0}}$ etc.
…etc.

**Step #4:** Subtracting a one-digit number from between 11 and 19 with a result less than 10.
$9 = 11 - \underline{\phantom{0}} \rightarrow 11 - 2 = \underline{\phantom{0}}$
$8 = 11 - \underline{\phantom{0}} \rightarrow 11 - 3 = \underline{\phantom{0}}$
$7 = 11 - \underline{\phantom{0}}$ etc.
$9 = 12 - \underline{\phantom{0}} \rightarrow 12 - 3 = \underline{\phantom{0}}$
$8 = 12 - \underline{\phantom{0}} \rightarrow 12 - \underline{\phantom{0}} = 8$
$7 = 12 - \underline{\phantom{0}}$ etc.
$9 = 13 - \underline{\phantom{0}} \rightarrow 13 - 4 = \underline{\phantom{0}}$
$8 = 13 - \underline{\phantom{0}} \rightarrow 13 - \underline{\phantom{0}} = 8$
$7 = 13 - \underline{\phantom{0}}$ etc.
…etc.

**Step #5:** Subtracting a one-digit number from between 21 and 24 with a result less than 20.
$19 = 21 - \underline{\phantom{0}} \rightarrow 21 - \underline{\phantom{0}} = 19$
$18 = 21 - \underline{\phantom{0}} \rightarrow 21 - 3 = \underline{\phantom{0}}$
$17 = 21 - \underline{\phantom{0}}$ etc.
$19 = 22 - \underline{\phantom{0}} \rightarrow 22 - 3 = \underline{\phantom{0}}$
$18 = 22 - \underline{\phantom{0}} \rightarrow 22 - \underline{\phantom{0}} = 18$
$17 = 22 - \underline{\phantom{0}}$ etc.
$19 = 23 - \underline{\phantom{0}} \rightarrow 23 - 4 = \underline{\phantom{0}}$
$18 = 23 - \underline{\phantom{0}} \rightarrow 23 - 5 = \underline{\phantom{0}}$
$17 = 23 - \underline{\phantom{0}}$ etc.
…etc.

**Step #6:** Finding the difference between two numbers (working with the number 20).
$20 - 11$ (What is the difference b/t 20 & 11?)
$20 - 12$ (What is the difference b/t 20 & 12?)
$20 - 13$ (What is the difference b/t 20 & 13?)
…etc.

**Step #7:** Finding the difference between two numbers (working between 21 and 24).
$21 - 11$ (What is the difference b/t 21 & 11?)
$21 - 12$ (What is the difference b/t 21 & 12?)
$21 - 13$, etc.
$22 - 11$ (What is the difference b/t 22 & 11?)
$22 - 12$ (What is the difference b/t 22 & 12?)
$22 - 13$, etc.
$23 - 11$ (What is the difference b/t 23 & 11?)
$23 - 12$ (What is the difference b/t 23 & 12?)
$23 - 13$, etc.
…etc.

# –Second Grade Math–

## *Overview of Child Development in Second Grade*

The second grader is in the middle of an important developmental phase between the ages of six and nine. It is a golden time where the student is still king in their own kingdom. The second grader is now livelier and more aware of what is happening around her. She may even become mischievous. Physically, the roundness of the early childhood years has disappeared. The mouth is closed and the student becomes less dreamy. The child's personality and character become more evident. We can help the children to overcome the one-sidedness of their character by telling fables and saint stories. The joy of learning is based on strong habits, rhythms, and songs. The creative forces of the child increase along with an ability to create more vivid inner pictures. The second grader's confidence grows and builds upon the foundation laid in grade one. However, the student still needs strong leadership from the teacher through a consistent and rhythmical approach to the lessons. An artistic approach is used to awaken the intellect.

With the math lessons, this is the year to introduce the times tables – all of them brought in an artistic way. The child is eager to show their individual abilities, such as the joy of finding missing numbers, or the discovery of geometrical patterns in the times tables. At this age, the child's memory forces are very strong, so it is an ideal time to begin to learn the arithmetic facts.

## *Curriculum Summary for Second Grade Math*

The world of numbers
- The children should become fluent with counting up to 100. They should be able to start anywhere and continue counting without any hesitation, and do this either counting forward or backward.
- The students should gradually become comfortable with numbers up to 1000 (and possibly higher).
- *Estimating*. We build up from first grade estimation and progress to more challenging estimations.

Place value. Place value is introduced and practiced. Be sure that all children really get this; this is an important step needed in order to deeply understand the world of numbers.

Addition and subtraction facts. By the end of the year, the class should have learned their addition facts (up to 18) and corresponding subtraction facts *by heart* (e.g., 9+8=17; 17-8=9).

The times/division tables
- Multiplication and division go together. The times/division tables are really the same thing.
- By the end of the year, the class should be comfortable with all of the times/division tables from 2 to 12, (e.g., for the 7's table: 7, 14, 21, 28, etc.). This requires daily, systematic work!

The four processes
- *Addition*. By the end of the year, the class should be comfortable with adding any two-digit number with any one-digit number (e.g., 57+6).
- *Subtraction*. By the end of the year, the class should be comfortable with subtracting any one-digit number from any two-digit number (e.g., 52–6), and also with subtracting two 2-digit numbers such that the answer is a one-digit number (e.g., 72–69).
- Be sure to read *A Step-by-Step Progression for the Arithmetic Facts* (in the appendix).
- The students must gain a deeper understanding of the concept of multiplication and division, such as:
    3 x 2 = __   asks: "Three groups of two make what?"
    12 ÷ 3 = __   asks: "How many groups of three fit into twelve?"
- By the end of the year, the children should be able to do all four processes (even alternating on the same page) and know the difference between the processes without help – but keep the problems simple.
- The introduction to both vertical addition (i.e., carrying) and vertical subtraction (i.e., borrowing) should wait until third grade. Therefore, all written work with the four processes should still be done in horizontal form (e.g., 9=5+4, 5+4=9).

Time orientation. The children should become comfortable with the days of the week, the months of the year, and terms like "tomorrow", "yesterday", "noon", "afternoon", "evening", "six o'clock", etc.

Bring wonder and awe to the world of numbers. *Geometrical patterns* with the 10-point circle or the 12-point circle help with this. (See more details under Block #1, below.)

Remediation. By the end of second grade, it is important to identify those students who have not developed a healthy sense of number and create a plan to remediate this problem. (See p59 in Schuberth's book for more details on this topic.)

## Second Grade Math

### *Recommended Reading for Second Grade Math*

- Henning Anderson, *Active Arithmetic*. AWSNA Publications.
- Herman von Baravalle, *The Teaching of Arithmetic and the Waldorf School Plan*. Waldorf School Monographs, 1967, third edition.
- Else Gottgens, *Joyful Recognition*.
- Else Gottgens, *Waldorf Education in Practice*.
- Dorothy Harrer, *Math Lessons for Elementary Grades*. AWSNA Publications.
- Christoph Jaffke, *Rhythms, Rhymes, Games and Songs for the Lower School*. Published by the Pedagogische Forschung Stelle, Der Bund der Freien Waldorf Schule in Stuttgart.
- Lipping Ma, *Knowing and Teaching Elementary Mathematics*. Lawrence Erlbaum Publ. 1919, London.
- Ernst Schuberth, *Teaching Mathematics for First and Second Grades in Waldorf Schools*. Rudolf Steiner College Press.
- John A. van de Walle, *Elementary and Middle School Mathematics*. Virginia Commonwealth University, Longman, NY.

## *Lessons and Topics for Second Grade Math*

Scheduling
The curriculum calls again for twelve weeks of math. The first math main lesson should fall around October, the second block around January, and the last block falls in the spring. Each block should be four weeks long, and there should be one week of math review at the end of the school year.

### Don't forget!!

- *Every day*, we should review the previous day's lesson.
- *Every day*, we should bring something new to the children.
- *Every day*, the students should practice the new material and selected review topics.
- *Every day*, we should ask the students to estimate something: "What do you think the answer will be?"
- *Every day*, we should ask this question at least once: "How did you do that?"

### Practice and review (of old material)

- In order for the children to learn a topic well, it needs to be systematically reviewed. Obviously, all previous topics cannot be reviewed every day. The teacher needs to decide which topics are the most important to review each day. Some of these topics may need to be reviewed and practiced nearly every day, even when the class is not in a math block – although, it should take no more than a total of 10 minutes. (Also see *Review and Practice* under *Principles of Waldorf Education* in the *Introduction,* above.)

The world of numbers
- *Counting* up to 100 fluently, and up to 1000 comfortably – forward and backward. Children should be comfortable with doing this on their own without help.
- *Number dictations*. Number dictations should be used to help the students become familiar with the larger numbers that come about through the introduction of place value. Dictations should continue after the place value block ends. Students need to feel confident in writing any number between 1-1000.
- *Language*. Be careful when speaking these new, larger numbers (and expect the same of the students), so that an extra "and" isn't added in. For example, 270 should be spoken as "two-hundred-seventy", not "two-hundred-*and*-seventy".[29]

Addition and subtraction facts. By the end of the year, the class should have learned their addition facts (up to 9+9=18) and corresponding subtraction facts *by heart*.

---

[29] This is given from the perspective of the USA. Other English-speaking countries may be different.

# Second Grade Math

<u>Strategies</u>. Practice using strategies for addition and subtraction. For example, 24 +13 can be looked at as 24+10 +3 or 20+10+4+3 or 24+6 +7, etc. It is wonderful for the students to share strategies with each other. Every day, we should ask this question at least once: "How did you do that?" We can be quite surprised to hear all the strategies that the students can think of!

<u>Times/division tables</u>
- Remember that times/division tables are the highest priority for math in second grade.
- *The transition from skip counting to the times tables*. In first grade, we did rhythmical counting and skip counting, but did not speak conceptually about the times tables. Now, in second grade, we make the transition from skip counting (e.g., 3, 6, 9, 12, etc.) to understanding it as a table (3=1x3, 6=2x3, 9=3x3, 12=4x3, etc.).
- After working with each table through movement exercises, the children should write it out – both in full form (3=1x3, 6=2x3, 9=3x3, 12=4x3, etc.), and in a row (3, 6, 9, 12, etc.), which is simply a written version of skip counting.
- Be sure to frequently also "translate" the table into spoken words. For example, "twenty-one is seven groups of three."
- The grand goal is that by the end of the year, the students are comfortable with writing out all of the times/division tables from 2-12 in a row (3, 6, 9, 12, etc.). This requires daily, systematic work!
- The times tables and the division tables are worked on at the same time. The students can then experience how multiplication and division are related.
- Read again the "Progression for Rhythmical Exercises" found in the Introduction under the *Principles of Waldorf Education*.

<u>Practice with the four processes</u>. The students should continue practicing the four processes daily and develop strategies to solve these problems. The students need to have sufficient practice time, so that they start to feel confident. Also, try to make the problems interesting. For example, you can present a sequence of problems that reveal a special pattern (as shown on the right). Did you notice that the last one is the "little dragon"?

$8+7 = 15$
$18+7 = 25$
$28+7 = 35$
$38+7 = 45$
$68+7 = 75$
$78+6 = 84$

<u>Time</u>. Once the days, months, seasons, and clock time have been covered in main lesson, they should all be reviewed at least briefly, so that the students remain fluent with these concepts.

<u>Mental arithmetic</u>
- Practice mental arithmetic daily (for 10 minutes or less).
- Mental arithmetic also helps the children to become flexible in using the four processes.
- Work up to calculations like  $28 + 5$; $14 \div 2$; $8 \times 5$; $86 - 3$; $76 - 74$; $42-6$, $72-69$.
- *Halfway problems*. These problems should start off easy in second grade and get increasingly more difficult in the coming years. Some examples are:
  - What number is halfway between 25 and 29?
  - What number is halfway between 5 and 13?
  - What number is halfway between 5 and 23?
  - What number is halfway between 25 and 31?

<u>Word Problems</u>. In second grade, word problems may still be done through imaginative stories, and they can also be done more simply during the mental arithmetic section of the lesson. Here are some examples:
- An empty bus stops and picks up 8 people. At the next stop, 3 people get off and then 4 get on. How many people are now on the bus?
- A mother went to the market with her two children. They bought 24 oranges. They looked so delicious and all three of them ate one orange right away. On their way home, the children met a friend and all three children ate another orange. How many oranges were in the mother's basket when they got home?

# Second Grade Math

- A father was working in the garden during the fall. He found a hole with 5 nuts in a tree, later he found 6 nuts buried in the ground, and then he found 8 nuts on a shelf in the garden shed. How many nuts were there in total?

Games. Try to find engaging games, including games that use the number line (see Henning Anderson's book). These can be wonderful activities for the children to do when there is extra time. (See the appendix for a list of games for lower school math.)

*Spotting Tens.* Here's a card game for grades 1 and 2:
- Materials: Use a normal deck of cards where the face cards and the 10's have been removed. The aces will count as a 1. Alternatively, you can use a deck of Krypto cards, or have the students make their own cards.
- How to play: Each student has a partner – one is the dealer, and the other is the player. The dealer puts seven cards face down on the table – one card in the center, and six cards around the center card. The objective is for the player to spot pairs of cards that add to 10. When the player spots a pair adding to 10, she removes that pair, and then the dealer replaces those two spots with new cards. After the deck is finished, the dealer and player switch roles. To make the game more challenging, you can look for combinations of cards (perhaps even 3 or 4 cards) that add to a larger sum, like 12 or 15.

Practice book. In second grade, it is good for the children to keep a practice book. What has been done through movement exercises can now be written down. A practice book is very handy and helps the children learn how to organize their work. The work in this practice book should still be done neatly.

---

## Math Main Lesson Block #1 – Place Value and the Times Tables

New material and content
- *The times tables*.
  - In this first math block of second grade, we begin to introduce the times/division tables.
  - Keep in mind that the goal is that by the end of the year, the class should be comfortable with all of the times tables from 2-12. This requires daily systematic work, and much careful planning.
  - Be sure to read *A Step-by-Step Progression for the Arithmetic Facts* (in the appendix).
- *Place value*. It is best to work with images when introducing place value.
  - We can create a story about dwarves who mine for gems. They need to find a way to keep track of the gems and count them. It may be good for the students to try first to come up with a system on their own. One idea is as follows: Groups of 10 gems are put into small pouches. Groups of 10 pouches are put into small buckets. Groups of 10 buckets are put into wheelbarrows, etc. In this way, the number 3,785 is represented by 3 wheelbarrows, 7 buckets, 8 pouches, and 5 (single) gems. This is a nice story to illustrate in a main lesson book. It can lead into the next (perhaps less elaborate) story...
  - *A woodcutter's mill*. A woodcutter has a lumber mill that produces boards for sale in the village. Groups of 10 boards get tied into bundles. Groups of 10 bundles get put into crates. Groups of 10 crates get put onto horse wagons. In this way, the number 2,497 is represented by two wagons, 4 crates, 9 bundles, seven (single) boards. This story nicely lends itself to the children practicing place value by using craft (popsicle) sticks, which are available in most craft stores. You can simply give the students a large number of sticks, maybe around 2,400 sticks, and have the students group them like the woodcutter so they can count all of the sticks.
  - After the introduction of place value, number dictations can help the students practice with these newly discovered larger numbers.
- *The four processes*. Develop strategies for addition and subtraction. The students should be fluent with sums up to 24, and comfortable up to 100. This work will continue throughout second grade.

Second Grade Math

Activities and Movement
- *Ten children on a circle.*
    - Ten children stand in a circle while the rest of the class observes. The first child speaks "one", the child on her left speaks "two", and the counting continues like this, clockwise around the circle, stopping finally at 60, or maybe further. We then ask one child in the circle: "What was the first number you said?" Perhaps, he says "three". And then we ask the class: "What was the next number he said?" They come to see the nice pattern that emerges – if we know the first number that someone says, then we know the rest (e.g., 3, 13, 23, 33, etc.).
    - The next day, we repeat the exercise once, and then do it one more time, but now we place one more child in the circle, and repeat the process. The pattern is a bit trickier, but exciting for the children to discover! That same third child now has these numbers: 3, 14, 25, 36, etc.
    - There are many ways to play with this activity and create something new and exciting for the children. (See Henning Andersen's book (p68-72) for more ideas related to this.)
- There are many ways to represent patterns of numbers in geometric form. Below, we present two possibilities for working with the times tables within a circle: the 10-circle and the 12-circle. We encourage the teacher to choose one option or the other. Doing both circles in second grade is not recommended because it would lead to unhelpful confusion.
- *The 12-circle.* All of the times tables can also come out of a 12-circle. We like to introduce the circle of twelve because twelve represents a cosmic reality. The children are still very connected to the cosmos, the world they come from. The students will recognize the twelve in the clock, and in the twelve months (second block). We can do this for all of the times tables (starting with the 2's times table). For example, consider the 5's table. The children sit on chairs in a circle, representing the numbers of the clock. Another child stands in the center of the circle with a ball of yarn. The child representing the number 12 (which is also zero) holds one end of the yarn. The child in the center takes 5 steps and ends up at the number 5. This child takes the string and everyone says: "five equals one times five." In this way, the middle child passes through the numbers of the 5's times table all the way up to 60=12x5. The yarn went around the circle several times. Once this is completed, we have woven together a beautiful 12-pointed star.

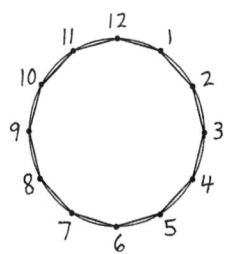

**1's and 11's Table**

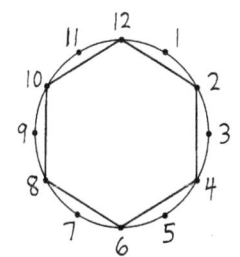

**2's and 10's Table**

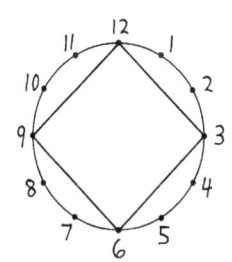

**3's and 9's Table**

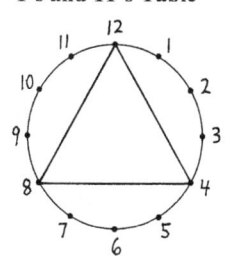

**4's and 8's Table**

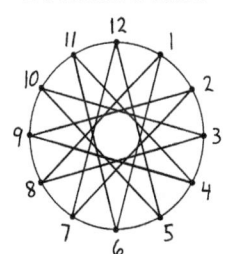

**5's and 7's Table**

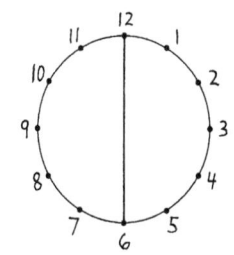

**6's Table**

# Second Grade Math

- *The 10-circle*
    - The whole above process (just described with the 12-circle) can also be done with the 10-circle. In this case, we start with a 10-pointed circle, where each child on the circle represents the digits zero through nine. The pattern we now get with the yarn follows the last digit of the chosen times table.
    - For example, using the 4's table, the yarn is passed from the 4 (**4**=1x4), to the 8 (**8**=2x4), to the 2 (because in this third step, 1**2**=3x4, which has 2 as a last digit), to the 6 (1**6**=4x4), to the 0 (2**0**=5x4), and then the digits repeat so the yarn revisits the same five children: 2**4**=6x4; 2**8**=7x4; 3**2**=8x4; 3**6**=9x4; 4**0**=10x4; 4**4**=11x4; and finally 4**8**=12x4. We have created a five-pointed star, which keeps repeating, again and again, no matter how far we choose to go – even if we went up to 100=25x4 we would get five 5-pointed stars laid on top of one another!

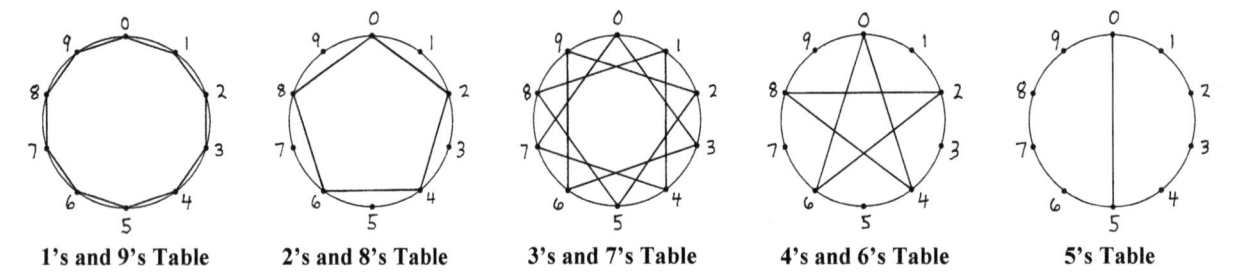

| 1's and 9's Table | 2's and 8's Table | 3's and 7's Table | 4's and 6's Table | 5's Table |

- Why do the 4's and 6's tables produce the same geometrical pattern? Because they have similar qualities! The last digit for the 4's table cycles every five steps (4,8,2,6,0), and the 6's table has the same cycle, but backwards (6,2,8,4,0). In the same way, the 1's table and the 9's table both produce the same 10-sided figure (decagon); the 2's table and the 8's table both produce pentagons; and the 3's table and the 7's table both produce beautiful 10-pointed stars.
- Of course, all of this can be quite an amazing discovery for the second grader! They will come to the realization that for any two numbers that add to ten, their tables will produce the same geometric pattern, but it is produced backwards. This is because the last digits in their tables follow the same pattern.
- *Stepping.* Start at any given number between 100 and 1000 and have the whole class count forward and backward. Then do this in small groups and individually.
- *Clapping and stamping – Important!* Practicing times tables through rhythmical exercises.
    - For example, with the 4's times table, each child has a partner and does the following: **_four_** (quarter note, clap hands with partner) *is* (quarter note, clap your thighs) *one* (eighth note, clap your own hands) *times* (eighth note, arms crossed) *four* (quarter note, snap fingers); **_eight_** (quarter note, clap hands with partner) *is* (quarter note, clap your thighs) *two* (eighth note, clap your own hands) *times* (eighth note, arms crossed) *four* (quarter note, snap fingers); **_twelve_** (quarter note, clap hands with partner) *is* (quarter note, clap your thighs) *three* (eighth note, clap your own hands) *times* (eighth note, arms crossed) *four* (quarter note, snap fingers), etc.
    - However, be careful with the above exercise – sometimes teachers do this with the whole class so often, and with no variation, that it becomes like a church drone; the children become bored, and aren't conscious of what they are saying. We need to find ways to make it interesting and bring it into their consciousness. (See *Progression for Rhythmical Exercises*, under *Principles for Waldorf Education* in the Introduction.)
    - One way to vary the above exercise is to only speak the numbers in the table. So, the clapping rhythm is the same, but you only speak the underlined words 4, 8, 12, 16, etc., with a bit of silence between each spoken number.
    - There are so many possibilities for rhythmical exercises to practice times tables. See what new exercises you and the class can create!
- *More ideas for movement.* For more movement ideas see Henning Andersen's book, *Active Arithmetic.*

# Second Grade Math

<u>Working with manipulatives</u>
- Manipulatives should be used less frequently in second grade than in first grade, but likely there are a few children who will still need the manipulatives (e.g., a string with beads).
- The number line can be used for games and activities. We can ask questions like: Where is the number that is 5 more than 26?
- Board games can be fun, and some games with dice can help practice mental arithmetic. (See the appendix for a list of games for lower school math.)

<u>Bookwork</u>
- The beautiful geometrical forms we created during our movement exercises (with either the 12-circle or the 10-circle) can now be drawn into the children's main lesson books.
- Be sure to visit our website (www.JamieYorkPress.com). Under the "Resources" tab you can see full-color pages from students' main lesson books.

## Lesson Plan Outline for Block #1 – Place Value and the Times Tables
(Does not include all topics needing review from first grade)

Week #1
- Introduce place value through a story, and then act it out and practice it.
- Introduce the concept of times/division tables for the 2's table. This should include:
  - Movement exercises.
  - Writing it down both in full form (2=1x2, 4=2x2, etc.) and in a row (2, 4, 6, etc.)
  - Representing the 2's table in the 10-circle or the 12-circle.

Week #2
- Practice place value.
- Number dictations with large numbers.
- 3's and 4's times/division tables, including all of what was done with the 2's table (see above).

Week #3
- Practice place value and number dictations.
- 5's and 6's times/division tables, including all of what was done with the 2's table (see above).
- Work on new strategies with the four processes.

Week #4
- Wrap up the block.
- Keep practicing everything from weeks 1-3.
- 7's times/division table, including all of what was done with the 2's table (see above).

Second Grade Math

# Math Main Lesson Block #2 – Time and the Times Tables

New material and content
- *Time orientation.* The main focus in this block is time: the days of the week, the months, and the clock. The children learn to read a clock with hands, where the minute hand is straight up on the 12, such as 4 o'clock, etc. We add the minutes in third grade.
- *The times/division tables.*
  - Review the 2-7's times/division tables.
  - Introduce the 8's and 9's times/division tables.
  - Be sure to read *A Step-by-Step Progression for the Arithmetic Facts* (in the appendix).
- *The four processes.* We can slowly bring more difficult problems (but not too difficult!) to the children. Problems are presented both orally (with mental arithmetic) and in written form. Also, continue to develop strategies.
- *Place value.* We should review and further what was introduced in the first block.
- *Number dictations.* Number dictations should be done in order to help the students build an understanding of place value. With these number dictations we can vary the digits slightly, such as: 21, 12, 23, 32, 41, 14, 112, 121, 211, etc.
- *Making a calendar.* It may be best if this block falls in January, which is a perfect time for making calendars.
- In Christoph Jaffke's book, *Rhythms Rhymes Games and Songs for the Lower School*, one can find excellent poems for the theme of time and season.

Review and practice! Previously introduced material needs to be reviewed, practiced, and furthered. (See above, *Practice and review of old material under Lessons and Topics for Second Grade Math.*)

Activities and Movement
- *Movement exercises for the times tables.* Continue with the movement exercises for the times tables from block #1, and add some of your own – perhaps using rhythm sticks, beanbags, and more.
- *Movement exercises for place value.* The class teacher can find and invent a variety of games that practice place value.
- *Moving the four processes on the number line.* For example, if you start at 38, and step forward by ten, where do you end up? And then, if you take another step? We can also do this with subtraction. This can also be helpful to develop strategies. For example, instead of stepping forward by 13, we first could take a step of ten followed by a step of three.
- *Moving with the clock.* The clock can be visualized by making a circle with 12 students and then having two different students represent the hour hand and the minute hand.
- *Moving with the weeks and months.* We walk the days of the week and the months of the year forward and backwards with the class, in small groups, and individually.

Working with manipulatives
- *Making a clock.* The students can make a cardboard clock themselves, and then practice with it.

Bookwork
- The students continue to put the times/division tables in their main lesson books in an artistic way.
- *Pictures of time.* At 6 o'clock (morning) the sun comes up and we have breakfast; at 10 o'clock we have snack; at 12 o'clock (noon) the sun is overhead and we have lunch; at 6 o'clock the sun is setting as we eat, etc. The students can draw all of these pictures beautifully in their main lesson books.
- The students can make a calendar with one page for each month. We can provide them with a pre-printed grid, which they can glue to a piece of paper. Every page has a poem of the month, and also has a drawing and includes the birthdays of their classmates.
- Be sure to visit our website (www.JamieYorkPress.com). Under the "Resources" tab you can see full-color pages from students' main lesson books.

Second Grade Math

## Lesson Plan Outline for Block #2 – Time and the Times Tables
(Does not include all topics needing review from previous blocks and years)

Week #1
- Introduce the days of the week, the months, and the seasons, including movement exercises that integrate these themes.
- Review and practice all work with the times/division tables (2-7) from block #1.

Week #2
- Introduce the calendar! In this week we cover the 4 months of January to April.
- Every student starts making a birthday calendar.
- Introduce the clock.
- Continue practicing all work with the 2's through 7's times/division tables.

Week #3
- Continue with the calendar: May to August.
- Practice clock time with whole and half hour.
- Introduce the 8's times table, including:
  - Movement exercises.
  - Writing it down both in full form (8=1x8, 16=2x8, etc.) and in a row (8, 16, 24, etc.).
  - Representing the 8's table in the 10-circle or 12-circle.

Week #4
- Wrap up the block.
- Continue with the calendar: September to December.
- Practice clock time, including "quarter past" and "quarter to" the hour.
- Introduce the 9's times table, including all that was done with the 8's table, above.

---

## Math Main Lesson Block #3 – Bringing it all together!

New material and content
- The major purpose of this main lesson block is to bring everything together that has been introduced previously. Of course, it is always best to find new variations to previously introduced material. New games (see the appendix for a list of games for lower school math), riddles and/or puzzles can be added.
- *The times tables.*
  - Review the 2-9's times/division tables.
  - Introduce the 10's, 11's, and 12's times/division tables.
  - Be sure to read *A Step-by-Step Progression for the Arithmetic Facts* (in the appendix).
  - The children should create a times table square soon after all of the tables (2 through 12) have been introduced. This square should be done beautifully, with nice colors. The children can then try to find hidden secrets and patterns in the times table square.
- *The four processes.*
  - We continue to work with the four processes.
  - Ensure that the concepts of multiplication and division are clear.
  - Working with *Fact Families* is a nice way to bring all of this together.
- *Fact families.*
  - At the end of second grade, we can make the students even more consciously aware of the interrelationship between the four processes. Working with "fact families" is one way to do this.
  - The basic idea is this: every arithmetic fact has three other facts within its "family".
  - We can start with the question: "If I know that 13-5 = 8, what else do I know?" Answer: 13-8=5; 8+5=13; 5+8=13. These four facts form a "fact family". Likewise, the facts 5x3=15; 3x5=15; 15÷3=5; and 15÷5=3 form a fact family.
  - We can then ask the students questions like this: "What are the other members of the fact family 9+8=17?" Answer: 8+9=17; 17-9=8; 17-8=9.
  - Once the students have done place value, we can ask "big questions" like: "What is 5000 plus 3000?" Then we ask: "What are the other members of the fact family 5000+3000=8000?"

Review and practice! Previously introduced material needs to be reviewed, practiced, and furthered. (See above, *Practice and review of old material under Lessons and Topics for Second Grade Math.*)

# Second Grade Math

## Activities and Movement

- *Rhythmical exercises.* We should continue with daily rhythmical exercises, but we can now do more complicated movements for the times tables, and bring many variations. We can even *work with two different times tables simultaneously!*

- *Clapping and stamping.* We clap on the numbers that are in the 2's times table, and, at the same time, stamp our feet on the numbers that are in the 3's times table. All of this happens while we speak the numbers sequentially. (See below, where "c" = clap, and "s" = stamp.) We can also use different percussion instruments.

| 1 | 2 | 3 | 4 | 5 | 6 | 7 | 8 | 9 | 10 | 11 | 12 | 13 | 14 | 15 | 16 |
|---|---|---|---|---|---|---|---|---|----|----|----|----|----|----|----|
|   | c | s | c |   | cs |   | c | s | c  |    | cs |    | c  | s  | c  |

- *Stepping and hopping along a line with two tables.* Two children move together along a number line while holding hands – the one on the left stepping and hopping the 2's table, and the one on the right stepping and hopping the 3's table. Let the class observe when both of them hop at the same time.

- *Two tables inside circles*
    - This is an excellent way to show the relationship between two times tables.
    - This activity is an expansion of the movement exercise called "*Rhythmical/skip counting inside a circle*" that we did in first grade. (If you are not familiar with that exercise, then you will find it above, with first grade, block #2.)
    - You will first need to review *rhythmical/skip counting inside a circle* from first grade. Once the children have become comfortable with working with one table at a time, you can then do two tables simultaneously – then the real fun begins! The above drawing shows the 3's table on the left and the 4's table on the right. One child is the number runner for the 3's circle, and another child is the number runner for the 4's circle. They both start at "home" (shown above as "C"). The teacher counts slowly, as both runners move around their circles (at different rates), both in a clockwise direction. At the moment the teacher says "3", the runner on the 3's circle is back home, but the 4's runner is at F. Whenever the two runners reach home at the same time, they should clap hands together. The rest of the class is carefully watching – half of them speaking the numbers every time the 3's runner reaches home (3, 6, 9, 12, etc.), and the other half of the class speaking the numbers every time the 4's runner reaches home (4, 8, 12, 16, etc.). When we are finished – perhaps counting all the way to 60 – we ask the class when did the two runners clap hands? The class can then discover that the "clapping numbers" are the numbers in the 12's table!
    - Once the students have become proficient, we can choose different tables to work with each day. The children may be surprised to see that the "clapping numbers" are not always the product of the two tables we are working with. For example, with the 3's and 4's tables the clapping numbers are the 12's table, but with the 3's and 6's tables the clapping numbers are the 6's table, and with the 10's and 4's tables the clapping numbers are the 20's table.
    - This whole exercise sets the stage for common multiples and common denominators in fourth grade. It will be good to revisit this at that time.

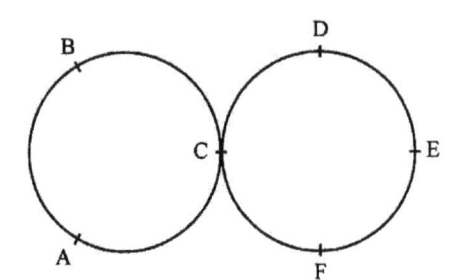

## Working with manipulatives

- Manipulatives are used less and less frequently. (See *Regarding number lines and manipulatives* under *The Art of Teaching Math* above.)

## Bookwork

- The children should finish their work with the times tables in their main lesson books.
- The children can now put the *times table square* into their book. In many ways, this is the culmination of all of our work through the year with the times tables. It should be done very beautifully!
- Be sure to visit our website (www.JamieYorkPress.com). Under the "Resources" tab you can see full-color pages from students' main lesson books.

Second Grade Math

# Lesson Plan Outline for Block #3 – Bringing it All Together!

(Does not include all topics needing review from previous blocks and years)

Throughout all four weeks of this block:
- Practice movement exercises for the times tables, and other exercises.
- Practice writing and speaking all of the times tables in a row – every day!
- Practice place value and number dictations.
- Practice strategies for the four processes.
- Practice time.

Week #1
- Introduce the 10's and 11's times table.

Week #2
- Introduce the 12's times table.
- Work with fact families.

Week #3
- Learn by heart all addition and subtraction facts up to 18 (e.g., 9+8=17; 17-8=9).

Week #4
- Practice different times tables simultaneously.

---

## *More Ideas for Teaching Second Grade Math*
(Be sure to reread relevant sections from the *Introduction*.)

Commutative property. Years ago, schools taught the commutative property (e.g., 4x3 is the same as 3x4) and also the associative property – both in a very abstract way. There is, of course, no real reason to have the children become familiar with these terms, but we do want to look for opportunities for the children to discover the principles behind this. For the children, 5x3 (i.e., five groups of three) is a very different process than 3x5 (i.e., three groups of five). Likewise, 2+6 is a different process than 6+2, yet both yield the same sum. This can be an exciting "aha" moment for the children when they discover for themselves that the two numbers being multiplied or added can be switched and the result will be the same.

Different kinds of subtraction
- *Three ways of looking at minus.* There are (at least) three ways to look at a subtraction problem. (This is for the teacher to keep in mind; the students may not be fully conscious of this until third grade.) Using the example of 20 – 13:
    - *As a Take Away (or Subtraction) problem.* We take away 13 from 20.
    - *As a Difference Problem.* We ask ourselves: "How far apart are 13 and 20?" or "How many steps are there going from 13 to 20?" This will be especially clear to the students once they have moved it on the number line.
    - *As the Reverse of Addition.* We ask ourselves: "13 plus what is 20?"
- The goal is to develop flexibility in the children's thinking, so that sometimes they look at it as a *difference* problem, and sometimes as a *take away* problem, depending upon which approach is easier for a given problem.
- The teacher should try to be consistent in the use of the words "take away", "difference", "subtract", etc. By saying "26 minus 14", we leave it open for the children to solve the problem by either *taking away* or *finding the difference*.
- The students should still be doing subtraction predominantly in the horizontal form (and in mental arithmetic) well into third grade, and beyond.

No "carrying" or "borrowing" yet. It is best to delay vertical addition and subtraction until third grade.
(See *Keep in the horizontal as much as possible* under *Developing a Sense of Number/Topics in the math classroom,* in the Introduction.)

# Second Grade Math

## A step-by-step progression for addition in second grade

**Notes:**
- Of course, there are more steps possible than are listed below. This is intended to give the teacher some possible ideas of what could be done to work in a structured way.
- Important: remember that this is all "mental arithmetic". Vertical addition isn't introduced until third grade.
- Because these problems aren't done using vertical addition (i.e., carrying), this all lends itself to developing strategies. For example, after we have done quite a bit of this work, imagine all of the ways that a student could figure out 46+19? Ask them to share their strategies with the class.
- The answer doesn't always have to appear at the end of the problem; it is important to vary the presentation of these problems, so that the students are flexible. For example, instead of $29+6=\_\_$, we could write it as $29+\_\_=35$.
- It is important to find "the story" behind a problem, or to "translate" the problem into spoken language. For example, with $7+\_\_=12$, I ask myself, "How much do I need to add to 7 in order to get 12?" or "Seven plus what makes twelve?", or "What is the difference between 12 and 7?"

**Step #1:** Adding multiples of 10.

$10+10=\_\_$ → $20=10+\_\_$
$20+10=\_\_$ → $30=20+\_\_$
$30+10=\_\_$ etc.

$20+20=\_\_$ → $40=20+\_\_$
$30+20=\_\_$ → $50=30+\_\_$
$40+20=\_\_$ etc.

$30+30=\_\_$ → $60=30+\_\_$
$40+30=\_\_$ → $70=40+\_\_$
$50+30=\_\_$ etc.

**Step #2:** Two-digit plus one-digit (without carry).

$25+3=\_\_$ → $28=\_\_+3$
$36+2=\_\_$ → $38=\_\_+2$
$72+5=\_\_$ → $77=\_\_+5$
etc.

**Step #3:** Adding 10.

$25+10=\_\_$ → $35=\_\_+10$
$36+10=\_\_$ → $46=\_\_+10$
$58+10=\_\_$ → $68=\_\_+10$
etc.

**Step #4:** Two-digit plus a multiple of 10.

$25+30=\_\_$ → $55=\_\_+30$
$36+40=\_\_$ → $76=\_\_+40$
$58+20=\_\_$ → $78=\_\_+20$
etc.

**Step #5:** Two-digit number plus a number between 10 and 20 (without carrying).

$35+13=\_\_$ → $48=35+\_\_$
$74+12=\_\_$ → $86=74+\_\_$
$52+16=\_\_$ → $68=52+\_\_$
etc.

**Step #6:** Two-digit number plus a number between 10 and 20, and with ones digits adding to 10.

$44+16=\_\_$ → $60=44+\_\_$
$67+13=\_\_$ → $80=67+\_\_$
$63+17=\_\_$ → $80=63+\_\_$
$81+19=\_\_$ → $100=81+\_\_$
etc.

**Step #7:** Two-digit doubling.

$24+24=\_\_$ → $48=\_\_+\_\_$
$32+32=\_\_$ → $64=\_\_+\_\_$
$45+45=\_\_$ → $90=\_\_+\_\_$
$49+49=\_\_$ → $98=\_\_+\_\_$
etc.

**Step #8:** Adding any two 2-digit numbers. The big, challenging step!

$53+18=\_\_$ → $71=53+\_\_$
$43+19=\_\_$ → $62=43+\_\_$
$26+54=\_\_$ → $80=26+\_\_$
$48+34=\_\_$ → $82=48+\_\_$
etc.

# Second Grade Math

## A step-by-step progression for subtraction in second grade

**Notes:**
- Of course, there are more steps possible than are listed below. This is intended to give the teacher some possible ideas of what could be done to work in a structured way.
- Important: remember that this is all "mental arithmetic". Vertical subtraction isn't introduced until third grade.
- Because these problems aren't done using vertical subtraction ("borrowing"), this all lends itself to developing strategies. For example, after we have done quite a bit of this work, imagine all of the ways that a student could figure out 42–15.
- The answer doesn't always have to appear at the end of the problem; it is important to vary the presentation of these problems, so that the students are flexible. For example, instead of 29 – 6 = __, we could write it as 29 – __ = 23, or __ – 6 = 23.
- It is important to find "the story" behind a problem, or to "translate" the problem into spoken language. For example, with 17=22–__, I can ask: "How many do I need to take away from 22 in order to get 17?"

**Step #1:** Subtracting multiples of 10.
20 – 10 = __  →  10 = 20 – __
70 – 30 = __  →  40 = 70 – __
100 – 60 = __  →  40 = 100 – __
etc.

**Step #2:** Subtracting one-digit from two-digits (*without* regrouping).
29 – 3 = __  →  26 = 29 – __
35 – 4 = __  →  31 = 35 – __
77 – 6 = __  →  71 = 77 – __
etc.

**Step #3:** Subtracting one-digit from two-digits (*with* regrouping).
21 – 3 = __  →  18 = __ – 3
35 – 8 = __  →  27 = __ – 8
77 – 9 = __  →  68 = __ – 9
etc.

**Step #4:** Subtracting two 2-digit numbers resulting in a one-digit number, *part I*.
29 – 23 = __  →  6 = 29 – __
35 – 31 = __  →  4 = 35 – __
77 – 72 = __  →  5 = 77 – __
etc.

**Step #5:** Subtracting two 2-digit numbers resulting in a one-digit number, *part II*.
33 – 26 = __  →  7 = 33 – __
44 – 38 = __  →  6 = 44 – __
77 – 68 = __  →  9 = 77 – __
etc.

**Step #6:** Subtracting multiples of ten.
33 – 10 = __  →  23 = 33 – __
59 – 20 = __  →  39 = 59 – __
87 – 40 = __  →  47 = __ – 40
etc.

**Step #7:** Subtracting 11 to 19 *without* regrouping.
48 – 15 = __  →  33 = 48 – __
56 – 12 = __  →  44 = 56 – __
97 – 16 = __  →  81 = 97 – __
etc.

**Step #8:** Subtracting 11 to 19 resulting in a multiple of ten.
48 – 18 = __  →  30 = 48 – __
56 – 16 = __  →  40 = 56 – __
97 – 17 = __  →  80 = __ – 17
etc.

**Step #9:** Subtracting 11 to 19 *with* regrouping – developing strategies!
53 – 18 = __  →  35 = 53 – __
(perhaps do 53-10-3-5, or 53–20+2)
82 – 14 = __  →  68 = 82 – __
47 – 18 = __  →  29 = 47 – __
etc.

**Step #10:** Subtracting any two 2-digit numbers. The hardest problems!
53 – 38 = __  →  15 = 53 – __
63 – 28 = __  →  35 = 63 – __
87 – 38 = __  →  49 = 87 – __
etc.

# –Third Grade Math–

## *Overview of Child Development in Third Grade*

The golden years come to an end as the child steps out of her little kingdom. This is the 9/10-year change, which usually comes towards the end of third grade. To the child, this can feel like a terrible loss and make her insecure. This is accompanied by a separation between the child and the world around her. She can also feel that her teacher has changed. The child may become lonely, insecure, and critical.

To help the children with this phase of their development we study the Old Testament. The lawfulness and lessons from the stories of the Hebrew people help the children feel more secure inwardly. Main lesson blocks like farming and house building help the children to develop a new relationship with the world around them.

The eagerness to learn is still strong; this is a time of real blossoming. With the sturdy foundation of the first two grades, the children can apply their mathematical knowledge to practical, everyday situations. Once the students have the feeling that they are nourished, they become less anxious about stepping into this new phase of their life. For the teaching of math, this means that we can bring them measurement and vertical arithmetic (e.g., carrying, borrowing, long multiplication). Having the children learn the arithmetic facts by heart is perhaps our most important goal.

## *Curriculum Summary for Third Grade Math*

The world of numbers
- The number world becomes fluent up to 1000 and comfortable with the numbers up into the millions.
- Place value work should be continued from second grade, but now with larger numbers.
- Number dictations should be continued in third grade. This can serve as an opportunity for "economical teaching". For example, we first give the numbers 546, 350, 306, 603, 218, 390 as a number dictation. Then we ask them to write them again, but this time vertically on the page from largest to smallest. Then we ask them to add the two middle numbers together. Then subtract the largest from the smallest. And lastly, as a challenge for those who need it, they can add the whole list together.

Learning all of the Arithmetic Facts!! **Very important!**
- Now is the time to learn the multiplication/division facts (that come from these tables) out of order. The groundwork was set in second grade, when the children learned their times/division tables (2-12). Also, all of the addition and subtraction facts are reviewed and strengthened.
- See *All about the Third Grade Arithmetic Facts Practice Sheets* (under *More Ideas for Teaching Third Grade Math*), and *A Step-by-Step Progression for the Arithmetic Facts* (in the appendix).

The four processes: Working vertically. Before third grade, the children have only been working with problems written horizontally. Now we introduce the vertical form.
- *Vertical addition.* This is the first experience with working vertically. Slowly, build up to adding two 4-digit numbers (e.g., 2346+4947). Vertical addition problems may be practiced extensively.
- *Vertical subtraction.* Build up to subtracting two 3-digit numbers (e.g., 643–387). Perhaps, less time is spent here than on vertical addition.
- *Vertical multiplication.* As with vertical subtraction, this could be just an introduction. In third grade, vertical multiplication should be limited to single digit multipliers (i.e., 2347 x 5 might be the hardest). Don't overdo it! Two-digit multipliers should wait until fourth grade.
- *Long Division.* The introduction to long division (i.e., 235÷5) should wait until fourth grade.

The four processes: Working horizontally.
- Although the children are introduced to working with the four processes in vertical form, most of their work with the four processes is still in horizontal form – as with their *arithmetic facts practice sheets*.
- We should begin practicing horizontal division problems that result in a remainder (e.g., 46÷7).

Measurement
- This is a fun, hands-on block!
- In this main lesson, we introduce time, distance, weight and volume.
- Additionally, bartering and the use of money is handled in a simple and practical way.
- Estimation is also an important part of this block.

Third Grade Math

## *Recommended Reading for Third Grade Math*

- Henning Anderson, *Active Arithmetic*. AWSNA Publications.
- Herman von Baravalle, *The Teaching of Arithmetic and the Waldorf School Plan*. Waldorf School Monographs, 1967, third edition.
- Else Gottgens, Joyful Recognition.
- Else Gottgens, Waldorf Education in Practice.
- Dorothy Harrer, *Math Lessons for Elementary Grades*. AWSNA Publications.
- Christoph Jaffke, *Rhythms, Rhymes, Games and Songs for the Lower School*. Published by the Pedagogische Forschung Stelle, Der Bund der Freien Waldorf Schule in Stuttgart.
- Lipping Ma, *Knowing and Teaching Elementary Mathematics*. Lawrence Erlbaum Assoc. Publishers 1919, London.
- Ernst Schuberth, *Teaching Mathematics for First and Second Grades in Waldorf Schools*. Rudolf Steiner College Press. (Especially the section about math weaknesses, pages 59 and up.)
- John A. van de Walle, *Elementary and Middle School Mathematics*. Virginia Commonwealth University, Longman, NY.

## *Lessons and Topics for Third Grade Math*

Scheduling
The curriculum calls again for twelve weeks of math (three main lesson blocks of four weeks each), with one week of review at the end of the school year.

### Don't forget!!

- *Every day*, we should review the previous day's lesson.
- *Every day*, we should bring something new to the children.
- *Every day*, the students should practice the new material and selected review topics.
- *Every day*, we should ask the students to estimate something: "What do you think the answer will be?"
- *Every day*, we should ask this question at least once: "How did you do that?"

### Practice and review (of old material)

- In order for the children to learn a topic well, it needs to be systematically reviewed. Obviously, all previous topics cannot be reviewed every day. The teacher needs to decide which topics are the most important to review each day. Some of these topics may need to be reviewed and practiced nearly every day, whether the class is in a math block, or not. As a general guideline, the class should practice 30 minutes daily, when in a math block, and about 10 minutes daily (during main lesson) when not in a math block. (Also see *Review and Practice* under *Principles of Waldorf Education* in the *Introduction*.)

Daily practice sheets. *Learning the arithmetic facts by heart.*
- The groundwork was set in second grade when the children learned their times/division tables from 2 to 12 (e.g., for the 7's table: 7, 14, 21, 28, etc.). They must now learn the multiplication/division facts (that come from these tables) out of order, as well as all of the addition facts and subtraction facts.
- Remember that multiplication and division always go together. If we are working with a particular multiplication fact (e.g., 7x4=28) then we are working with the corresponding division fact (e.g., 28÷4=7 and 28÷7=4) at the same time. Additionally, facts should also be practiced in a variety of different ways, such as 7x__=28, and __÷4=7.
- Read *All about the Third Grade Arithmetic Facts Practice Sheets* (under *More Ideas for Teaching Third Grade Math*, below), and *A step-by-step progression for the arithmetic facts* in the appendix.

The times/division tables
- *Movement.* Continue and further the movement exercises that were introduced in second grade (see especially block #3 in second grade, above). For example, in second grade, we did the times table of two (with a clap) and three (with a stamp) while we counted. Now, in third grade we can add nodding our heads with the 4's times table. Henning Anderson gives several other good examples.

# Third Grade Math

- Every time we do a times table with movement exercises, we should also find a way of writing it into our main lesson books so that it shows the relationships.

Practice with the four processes
- Even though the children in third grade are being introduced to vertical arithmetic, they still should practice horizontal arithmetic (e.g., with their daily practice sheets). It is also important to keep up daily mental arithmetic (through eighth grade!). All of this helps ensure that their sense of number continues to develop and that they strengthen their ability to do simple calculations in their head.
- We should continue to encourage the children to find strategies when working with the four processes. Every day, we should ask this question at least once: "How did you do that?"
- We should begin to practice simple division problems that result in a remainder (e.g., $46 \div 7 = 6\, r\, 4$).
- *Fact Families* (which may have been introduced in second grade) are a good way to bring into consciousness the interrelationship between the four processes, and can be practiced occasionally in third grade. (See second grade, under "*More Ideas for Teaching Second Grade Math*".)

Mental arithmetic
- Practice mental arithmetic daily (for 10 minutes or less), and continue to develop strategies.
- Work up to calculations like $46 + 5$; $28 + 31$; $75 - 68$; $73 - 4$; $24 \div 4$; $100 - 13$; $51 = 46 + \underline{\phantom{x}}$.
- *Halfway problems*. These problems should start off easy in second grade, and now, in third grade, can get a bit more challenging. Some examples are:
  - What number is halfway between 25 and 41?
  - What number is halfway between 25 and 61?
  - What number is halfway between 250 and 450?

Games. Simple puzzles, riddles, and games can be valuable at this age. (See *A list of Games* in the appendix.) Try to find engaging games, including dice and board games. These can be wonderful activities for the children to do when there is extra time.

*Summing Cards*. Here's a card game for grades 3 and 4:
- Materials: Use a normal deck of cards where the face cards have been removed. The aces will count as a 1. Alternatively, you can use a deck of Krypto cards, or have the students make their own cards.
- How to play: Each student has a partner – one is the dealer, and the other is the player. The dealer puts eight cards face down on the table – in two rows of four. The objective is for the player to add up all the cards as quickly as possible. Usually it is best for the player to remove pairs adding to ten as she is summing the cards together. Then at the end there is likely to be a few cards left (that don't have any pairs adding to ten). There can also be another student who serves as an observer, checking the work of the player. To make the game more challenging, you can begin instead with twelve cards, or more.

Reading clocks. In second grade, the students were introduced to reading clocks (with hands), at least as far as reading times like 4 o'clock, 7 o'clock, etc. (where the minute hand is up at 12). In third grade, we review and practice those things from second grade, and then we learn to read all possible times, such as 5:23, 11:04, etc. We also cover the following: 4:30 is the same as "half-past four"; 8:45 is "quarter to nine"; 10:50 is "ten to eleven"; and 7:25 is about "half-past seven".

Practice book. In second grade, the children began to keep a practice book, which should be continued in third grade, and beyond. As always, the work in a practice book should be done neatly. In third grade it teaches the children how to organize their work.

The world of numbers. We review counting up to 1000, and beyond, forward and backward. Number dictations should also be continued.

Third Grade Math

# Math Main Lesson Block #1 – Vertical Addition and Subtraction

New material and content
- *Vertical addition and subtraction* (a.k.a., "carrying and borrowing"). This is a new (and exciting!) subject.
    - This is just an introduction to vertical addition and subtraction. Over the course of the third grade year, we will build up to adding two 4-digit numbers (e.g., 8364+8375) and subtracting two 3-digit numbers (e.g., 643–387).
- *Terminology.* Try not to use the word "borrowing" with the students because we never actually give anything back in the process.[30] Instead, we should call the whole procedure "vertical subtraction", and the strategy that gets us around the tricky situation (where, in a given column, the top digit is smaller than the bottom digit) we will call "regrouping".
- The students first need to be confident with place value. Therefore, a strong review of second grade place value will be necessary.
- It is wise to also review the regrouping of numbers.
    - For example, how can we regroup 53 as the sum of two numbers? It could be written as 53 = 50+3, or 53 = 40+13, or 53 = 30+23, or 53 = 52+1, or 53 = 48+5, etc.
- Be sure to read *A step-by-step progression for vertical addition and subtraction* under *More Ideas for Teaching Third Grade Math*, below.

Review and practice! Previously introduced material needs to be reviewed, practiced, and furthered. (See above, *Practice and review of old material under Lessons and Topics for Third Grade Math*.)

Activities and Movement
- The times tables should now be done with more challenging movement. Different times tables could be "moved" at the same time.
    - *Example #2*: We can revisit the exercise "*Two tables inside circles*" (see block #3 of second grade).
    - *Example #2*: We can all clap the two, stamp the three, and nod your head on the four (or six).
    - *Example #3*: With the 3, 6, 9, and 12 tables, have one student count, perhaps by beating a drum, and assign the rest of the class different tables (3, 6, 9, 12). Different movements are assigned to each table. Perhaps the phlegmatics should have the 3's table, so that they move the most.
    - After doing such exercises, we should ask the class some questions, such as: When did we all move at the same time? When did only the 3's move? When was there no movement? etc.
    - Try to find many more ways of bringing movement to the times tables so that the class can work together in a joyful way.

Working with manipulatives
- Students should now be able to imagine the regrouping of the numbers, therefore manipulatives should be avoided.

Bookwork
- The students should add many vertical addition and subtraction problems into their books.
- The work needs to be neat and orderly! Use appropriate colors (e.g., green for adding, and blue for subtracting, etc.).
- Be sure to visit our website (www.JamieYorkPress.com). Under the "Resources" tab you can see full-color pages from students' main lesson books.

---

[30] Lipping Ma, *Knowing and Teaching Elementary Mathematics*.

Third Grade Math

# Lesson Plan Outline for Block #1 – Vertical Addition and Subtraction

(Does not include all topics needing review from previous blocks and years)

Week #1
- Introduction to vertical addition.
- Review 2's and 3's tables, then practice the 2's and 3's times/division facts – out of order!
- Practice horizontal addition and subtraction (oral and written) and develop strategies for problems up to 1000 (e.g., 780+50, and 500-70), which continues through the year.
- Integrate the "5 facts of the week" into the lessons, and start using the *Arithmetic Facts Practice Sheets*. (This work continues for the rest of the block, and the rest of the year.)

Week #2
- Introduction to vertical subtraction.
- Review 4's and 5's tables, then practice the 4's and 5's times/division facts – out of order!

Week #3
- Practice vertically adding and subtracting.
- Review 6's table, then practice the 6's times/division facts – out of order!

Week #4
- Practice vertically adding and subtracting.
- Review 7's table, then practice the 7's times/division facts – out of order!

---

# Math Main Lesson Block #2 – Measurement

New material and content
- *Regarding the US measurement system.* No doubt, the US measurement system is rather archaic, and requires tedious calculations. It is indeed quite strange how slow the country has been to fully embrace the metric system. Yet, from a pedagogical standpoint, there are certain advantages to the US measurement system, namely: (1) it comes more naturally from the human body; (2) discovery can easily be woven into the block (see below); (3) we can wait until fifth grade to introduce the metric system and integrate it with the introduction to decimals; (4) in seventh and eighth grade, the students can work with two different measurement systems, which leads to some excellent mathematics.
- *For those people not in the USA.* Even if the metric system is the standard, it may still be best to try to introduce linear measurement out of the human body (see below), and have weight and volume also arise from common objects (e.g., stones and cups). Likely, you can find out about the measurement system that existed in your country before the metric system came along. However, you should be using the metric system in the second half of this block so that the students become familiar with the how things are measured in their country today.
- *Discovery!* Rather than giving all the conversion factors (e.g., 12 inches = 1 foot) to the students, we can allow them to discover these. Have them figure out how many cups are in a pint, and how many quarts are in a gallon. For discovering the conversion factors for length, break the students into small groups, and give each group the following: (1) several toothpicks that have been broken into one-inch-long pieces; (2) four straight sticks that are one-foot long; and (3) one blank yard stick. They can then discover how many inches are in a foot, and how many feet are in a yard. Then they draw all the markings on their own yardstick. And if you really want to do something adventurous, go on a hike and have the students discover how many feet are in a mile! How? Before hiking with the students, mark out one mile. Give each group a 100-foot-long rope which is marked every 10 feet. The students then use the rope to measure the mile. Hopefully they get between 50 and 55 rope-lengths – around 5300 ft.
- *Measurement topics*.
  - *Linear measurement*.
    - The children are introduced to linear measurement based on the human body – e.g., a foot is the size of the king's (rather large) foot; an inch is the length of a bone in the hand, etc.
    - Many hands-on experiences! They should measure things in and around the classroom.
    - As a class, we can make a simple graph that shows one particular student's height over a period of several months – or maybe each child could do that at home.
  - *Weight*. Weight should also be part of the experience. The students can weigh objects in the classroom, and they can weigh one another.
  - *Volume and liquid measure*. With liquid measure we should pour cups into a pints, quarts, and gallons, and tablespoons into a cup, etc.

## Third Grade Math

- *Origins.* The historical origins of the various units of measurement can be brought in an imaginative way (e.g., through the story of Noah's Ark).
- *Practical use.* Of course, it is best to show practical uses of measurement, such as with simple building projects, cooking recipes, etc.
- *Estimation.* The students should practice estimating. For example, they could estimate the length of a board, the height of a tree, the distance between two rocks, the weight of an animal. We should always try to estimate before taking an actual measurement.
- *Unit conversions.* The students should practice very simple unit conversions, such as converting feet to yards, feet to inches, pints to quarts, etc.
- *Money.* This is also the time to introduce the children to our currency. They can make their own money (coins and bills) and stock a store, or even create a market place. We can also create games for money. Through all of this activity, the children practice counting money and making change.
- *Vertical addition and subtraction.* Vertical addition and subtraction were introduced in the first math block of third grade. In the meantime, this new material has been "put to sleep" (see *the conscious use of forgetting* under *Principles of Waldorf Education* in the *Introduction*). It is important that we not touch upon new material in between two math blocks. Now that the material (vertical addition and subtraction) has had its proper "rest", we can review it and then deepen it.
- *Division with remainders.* This should be done using the same dividend but different divisors. For example, using the number 14, divide it by 1, 2, 3, etc. (14÷1 = __; 14÷2 = __; 14÷3 = __; 14÷4 = __, etc.) After some time, the students will improve at figuring out the remainder. This is preparation for long division.

<u>Review and practice!</u> Previously introduced material needs to be reviewed, practiced, and furthered. (See above, *Practice and review of old material* under *Lessons and Topics for Third Grade Math*.)

<u>Activities and Movement.</u> Continue the movement activities that were done in the first block.

<u>Working with manipulatives</u>
- Of course, there are unlimited possibilities for hands-on experiences and activities during this measurement block. This should bring a lively spirit to the lessons and enthuse the students.
- Many props and visual aids are helpful for this block. For example, with volumes we could use milk cartons, jugs, bottles, spoons, cups, cans, etc.
- Our work with money and currency lends itself nicely to working with manipulatives.

<u>Bookwork</u>
- The children should write what they have experienced (e.g., one gallon is 4 quarts, etc.) in their main lesson books. They can make drawings (e.g., of a foot) as it applies to the units of measurement.
- Be sure to visit our website (www.JamieYorkPress.com). Under the "Resources" tab you can see full-color pages from students' main lesson books.

## Lesson Plan Outline for Block #2 – Measurement
(Does not include all topics needing review from previous blocks and years)

<u>Week #1</u>
- Intro to linear measurement: inch, ft, mile, etc.
- Review 8's table, then practice the 8's times/division facts – out of order!
- Continue *Arithmetic Facts Practice Sheets*, and practice horizontal problems and strategies, and continue through the rest of the block and year.

<u>Week #2</u>
- Introduction to weight: ounce, pound, ton.
- Review 9's table, then practice the 9's times/division facts – out of order!

<u>Week #3</u>
- Introduction to volume measurement: cup, pint, quart, gallon, peck, bushel.
- Review 11's table, then practice the 11's times/division facts – out of order!

<u>Week #4</u>
- Wrap up the block.
- Introduction to money.
- Review 12's table, then practice the 12's times/division facts – out of order!

Third Grade Math

# Math Main Lesson Block #3 – Vertical Multiplication

New material and content
- *Vertical multiplication.*
  - Vertical multiplication can only be successful when the students feel confident with the times tables.
  - In third grade, vertical multiplication should be limited to single digit multipliers (i.e., 2347 x 5 might be the hardest). Don't overdo it! Two-digit multipliers should wait until fourth grade.
  - Be sure to read *A step-by-step progression for vertical multiplication* under *More Ideas for Teaching Third Grade Math,* below.
- *Measurement practice.* Now that the students have experienced the measurement main lesson (in the previous block), we can practice some simple problems – but not too much. This also gives an opportunity to integrate the four processes with measurement. Here are some examples:
  - Li ran around a lake eight times. If one time around the lake is two miles long, how far did Li run in total?
  - Kate used 3000 bricks to build her house. If each brick weighs 20 pounds, what is the total weight of all the bricks? (Challenge: How many tons is this?)
  - Ben cut a 24-foot-long rope into 3 equal pieces. How long is each piece?
    or, how many 3-foot-long pieces can be cut from a 24-foot-long rope?
    (Note how the above two questions relate to the *three types of questions for multiplication/division* as we discussed in the first block for first grade (see first grade, block #1). The above questions are equivalent to these: How many are there in each group if we divide 24 into 3 equal parts? and How many groups of 3 are found in 24?)

Review and practice! Previously introduced material needs to be reviewed, practiced, and furthered. (See above, *Practice and review of old material* under *Lessons and Topics for Third Grade Math.*)

Activities and Movement
- It is a challenging but important task to bring all the times tables together in movement. We should continue and further these kind of exercises introduced in previous blocks. Here is another idea of how it could be done where the children are the workers on a ship:
  - The captain counts the 1's; the 2's are sweeping the deck; the 3's throw the buckets of water on the deck; the 4's raise the sails; the 5's pull up the anchor, etc.
  - Groups move and speak when they hear a number from their times table.
  - It is fine if some students only observe, since there are several questions that can be asked at the end of the exercise (e.g., When did we hear only the captain? When was the whole crew moving and speaking?, etc.).

Working with manipulatives
- We recommend keeping away from manipulatives during this vertical multiplication main lesson. The students need to internalize the processes.

Bookwork
- The children can include several examples (some with the steps) of vertical multiplication in their main lesson book.
- Be sure to visit our website (www.JamieYorkPress.com). Under the "Resources" tab you can see full-color pages from students' main lesson books.

Third Grade Math

## Lesson Plan Outline for Block #3 – Vertical Multiplication
(Does not include all topics needing review from previous blocks and years)

Week #1
- Introduction to vertical multiplication.
- Continue through the whole block:
  - Practice all the tables times/division facts out of order (2's through the 12's table).
  - *Arithmetic Facts Practice Sheets.*

Week #2
- Continue with vertical multiplication.
- Practice adding and subtracting vertically with larger numbers, especially numbers with 0's.
- Practice measurement.

Week #3
- Introduction to (mental) division with remainders – e.g., $14 \div 3 = 4\,r\,2$.
- Practice vertical addition, subtraction, and multiplication.
- Develop strategies for estimating horizontally with larger numbers.

Week #4
- Wrap up the block.
- Practice everything from the first three weeks.
- Work on confidence!

---

## *More Ideas for Teaching Third Grade Math*
(Be sure to reread relevant sections from the *Introduction*.)

### A step-by-step progression for vertical addition and subtraction

**Step #1:** Two-digit addition/subtraction without carrying or borrowing. (Many of these problems should be done during the first math block of third grade.)

$$\begin{array}{rcl} 35 & \to & 30+5 \\ +14 & \to & +(10+4) \\ \hline 49 & \leftarrow & 40+9 \end{array} \qquad \begin{array}{rcl} 56 & \to & 50+6 \\ -23 & \to & -(20+3) \\ \hline 33 & \leftarrow & 30+3 \end{array}$$

**Step #2:** Using and showing regrouping strategies. (Many of these problems should be done during the first math block of third grade.)

$$\begin{array}{rcl} 46 & \to & 40+6 \\ +38 & \to & +(30+8) \\ \hline 84 & \leftarrow & 70+14 \end{array} \qquad \begin{array}{rcl} 83 & \to & 70+13 \\ -25 & \to & -(20+5) \\ \hline 58 & \leftarrow & 50+8 \end{array}$$

**Note:** It is important that the children have worked a good deal with regrouping strategies (but not with vertical arithmetic) beginning in second grade.

**Step #3:** The shortcut – standard, vertical addition and subtraction. (Many of these problems should be done during the second math block of third grade.)

Using the same examples as above:

$$\begin{array}{r} \overset{1}{\phantom{0}}46 \\ +38 \\ \hline 84 \end{array} \qquad \begin{array}{r} {}^{7}\!8^{1}3 \\ -25 \\ \hline 58 \end{array}$$

**Notes:**
- In order to make these problems more meaningful, try creating problems that relate to various activities.
- It is best to introduce vertical addition and subtraction in an artistic way. For example:
  - With vertical addition, we can draw a house where there is a level for every number.
  - The answer ends up in the basement, and the attic is used for the "carry" numbers.
  - This should not be too elaborate of a story; it is simply an artistic way of writing it down.
  - Something similar can also be done for vertical subtraction.

## A step-by-step progression for vertical multiplication (with single-digit multipliers)

**Step #1:** Write down familiar multiplication facts in vertical form.

```
  4
 x7
 28
```

**Step #2:** A 2-digit number (in expanded form) times a single-digit.

```
 37  →   30 + 7      (The 2-digit number is
 x4  →    x  4        regrouped.)
          28
        +120
         148
```

**Step #3:** A 2-digit number (in expanded form) times a single-digit, with carrying.

```
         +20
 37  →   30 + 7
 x4  →    x  4
         140 + 8 = 148
```

**Step #4:** The shortcut – standard, vertical multiplication (again, 2-digit times 1-digit).

```
  2
 37
 x4
148
```

**Step #5:** A 3-digit number (in expanded form) times a single-digit number.

```
 486  →   400 + 80 + 6
 x 3  →           x 3
                   18
                  240
                +1200
                 1458
```

**Step #6:** A 3-digit number (in expanded form) times a single-digit, with carrying.

```
            +200  +10
 486  →   400 + 80 + 6
 x 3  →           x 3
          1400 + 50 + 8 = 1458
```

**Step #7:** The shortcut – standard, vertical multiplication (again, 3-digit times 1-digit).

```
 2 1
 486
 x 3
1458
```

**Step #8:** Don't write down the carry digit!

```
 486
 x 3
1458
```

It is important for the students to work towards not writing down the carry digit (i.e., holding the carry digit in their head).

Third Grade Math

## All about the *Third Grade Arithmetic Facts Practice* Sheets

- Free download!! You can download our *third grade arithmetic facts practice sheets* for free from our website: www.JamieYorkPress.com.
- Facts of the Week! (See next page.) The central idea of these sheets is that there are five facts of the week written on the board, which the teacher works on with the whole class during the week in a variety of ways (e.g., using movement, rhythmical work, games, etc.). These facts of the week then appear multiple times on the sheets for the current week, and are then reviewed systematically for the next several weeks.
- Background work. These sheets should ideally be the culmination of two years of work. If the work in first and second grade has been effective, then the children should feel that these sheets are easy. If these sheets become too difficult and tedious, then it is likely that the classroom work being done in preparation for these sheets is either insufficient or not effective enough.
- Timing and rhythm. The intention is that the first "third grade arithmetic facts" practice sheet should be done at some point between the end of September and the end of October in third grade. It can, of course, vary depending upon the class. After that, a sheet should be done (nearly) every day until the whole set of 100 sheets is completed. There are 20 weeks (100 days) of sheets in this set. Each sheet has 30 problems.
- Completion. Since these sheets are designed to be an integral part of learning the math facts, it is important that each sheet of the entire set be completed, otherwise certain facts won't get adequate exposure.
- Caution! This should be fun and easy for the students. If successful, this builds their confidence. It is important to make sure that these sheets don't become torture for the students. Try to de-emphasize the importance of speed. Help the students to realize that improvement is what is important.
- The whole picture. The 30 problems listed on a particular sheet are only a part of the daily math practice. **It would be very unfortunate if daily math practice consisted of nothing more than the 30 arithmetic facts practice problems that appear on these sheets.**
  - When the class *is not* in a math main lesson, daily math practice should take about 10 minutes. The 30 arithmetic facts may take 3-5 minutes. The remaining 5-7 minutes can be spent doing a few written (vertical) arithmetic problems, some brief rhythmical work, or something else.
  - When the class *is* in a math main lesson, daily math practice should take about 30 minutes. The 30 arithmetic facts may take 3-5 minutes. After that, the remaining 25 minutes of math practice (during a math main lesson) consists of the material that was brought in previous blocks and the current block.
- The elements of math practice. The following list shows some of the aspects to consider when planning math practice for the day.
  - *Arithmetic facts practice sheet*. The teacher copies by hand the 30 problems from our *Third Grade Arithmetic Facts Practice Sheets* onto paper to be photocopied.[31] (about 5 minutes)
  - *Mental arithmetic*. The teacher may decide to read the first 6 problems out loud.
  - *Extra math practice problems*. There should be a few extra problems that the teacher comes up with and writes on the board. The students copy them into their practice books and work out the answers. These problems also include practice and review of material covered in previous math blocks. (Takes 20-25 minutes if the class is in a math block, otherwise only 5 minutes.)
  - *Challenge problems*. It is important that the last few problems (that the teacher adds on) be more challenging in order to keep the "quicker" students fully engaged.
- What comes next? The next set of practice sheets is titled *Fourth Grade Arithmetic Facts Review Sheets* and is intended to thoroughly review the math facts covered on the first practice sheets.
- The hope is that just five minutes per day of practicing these arithmetic facts results in the whole class quite effortlessly learning their math facts by heart.
- And what happens if…? We hope that it won't happen, but there may be a few children at the end of third grade who still haven't solidly learned their arithmetic facts. In order to help these children, we can give them a multiplication/division table (i.e., a square for the tables). Additionally, these students should study the basic arithmetic problems with flashcards during morning practice time.

---

[31] *Saving paper*. Try to find ways to reduce the amount of paper being used. For example, rather than using one sheet of paper each day, each side could be divided into three columns (i.e., 1 sheet = 6 days). This is one of many ways to help develop an environmental consciousness in the students.

Third Grade Math

# The 105 Key Arithmetic Facts (each appears as a "fact of the week")

| | | | | | |
|---|---|---|---|---|---|
| 8 + 2 | 6 + 6 | 10 − 8 | 13 − 9 | 3 x 3 | 6 x 6 |
| 9 + 2 | 7 + 6 | 10 − 7 | 13 − 8 | 3 x 4 | 6 x 7 |
| 7 + 3 | 8 + 6 | 10 − 6 | 13 − 7 | 3 x 5 | 6 x 8 |
| 8 + 3 | 9 + 6 | 10 − 5 | 13 − 6 | 3 x 6 | 6 x 9 |
| 9 + 3 | 7 + 7 | 10 − 4 | 13 − 5 | 3 x 7 | 6 x 12 |
| 6 + 4 | 8 + 7 | 10 − 3 | 13 − 4 | 3 x 8 | 7 x 7 |
| 7 + 4 | 9 + 7 | 10 − 2 | 14 − 9 | 3 x 9 | 7 x 8 |
| 8 + 4 | 8 + 8 | 11 − 9 | 14 − 8 | 3 x 12 | 7 x 9 |
| 9 + 4 | 9 + 8 | 11 − 8 | 14 − 7 | 4 x 4 | 7 x 12 |
| 5 + 5 | 9 + 9 | 11 − 7 | 14 − 6 | 4 x 5 | 8 x 8 |
| 6 + 5 | | 11 − 6 | 14 − 5 | 4 x 6 | 8 x 9 |
| 7 + 5 | | 11 − 5 | 15 − 9 | 4 x 7 | 8 x 12 |
| 8 + 5 | | 11 − 4 | 15 − 8 | 4 x 8 | 9 x 9 |
| 9 + 5 | | 11 − 3 | 15 − 7 | 4 x 9 | 9 x 12 |
| | | 11 − 2 | 15 − 6 | 4 x 12 | 11 x 11 |
| | | 12 − 9 | 16 − 9 | 5 x 5 | 11 x 12 |
| | | 12 − 8 | 16 − 8 | 5 x 6 | 12 x 12 |
| | | 12 − 7 | 16 − 7 | 5 x 7 | |
| | | 12 − 6 | 17 − 9 | 5 x 8 | |
| | | 12 − 5 | 17 − 8 | 5 x 9 | |
| | | 12 − 4 | 18 − 9 | 5 x 12 | |
| | | 12 − 3 | | | |

# Facts of the Week

Week #1:   8 + 2;  7 + 3;  6 + 4;  9 + 2;  9 + 3;  9 + 4;  9 + 5;  9 + 6;  9 + 7;  9 + 8

Week #2:   5 + 5;  6 + 6;  7 + 7;  8 + 8;  9 + 9

Week #3:   10 − 8;  10 − 7;  10 − 6;  3 x 3;  3 x 4

Week #4:   8 + 3;  7 + 4;  6 + 5;  10 − 5;  3 x 8

Week #5:   8 + 4;  7 + 5;  10 − 4;  10 − 3;  10 − 2

Week #6:   8 + 5;  7 + 6;  11 − 9;  12 − 9;  13 − 9

Week #7:   8 + 6;  8 + 7;  14 − 9;  15 − 9;  3 x 7

Week #8:   16 − 9;  17 − 9;  18 − 9;  3 x 9;  4 x 5

Week #9:   11 − 8;  13 − 8;  3 x 6;  4 x 4;  5 x 5

Week #10:  11 − 4;  12 − 8;  13 − 4;  3 x 5;  5 x 6

Week #11:  11 − 7;  12 − 5;  16 − 7;  3 x 12;  5 x 8

Week #12:  12 − 3;  13 − 7;  14 − 7;  4 x 7;  5 x 7

Week #13:  11 − 5;  13 − 5;  17 − 8;  4 x 9;  5 x 12

Week #14:  12 − 4;  14 − 6;  15 − 6;  4 x 8;  5 x 9

Week #15:  11 − 6;  13 − 6;  14 − 5;  7 x 8;  4 x 6

Week #16:  11 − 2;  12 − 6;  15 − 7;  6 x 9;  6 x 12

Week #17:  11 − 3;  14 − 8;  15 − 8;  6 x 7;  6 x 6

Week #18:  12 − 7;  16 − 8;  6 x 8;  7 x 7;  7 x 9

Week #19:  4 x 12;  8 x 8;  8 x 9;  11 x 11;  9 x 9

Week #20:  8 x 12;  9 x 12;  7 x 12;  11 x 12;  12 x 12

# –Fourth Grade Math–

## *Overview of Child Development in Fourth Grade*

The gates of paradise have closed; the child has arrived on the earth. Steiner referred to this age (fourth and fifth grade) as the "heart of childhood". In fourth grade, the children start to find their way in the world. Their interactions with peers and adults can be challenging. They want to know more about the world, about personalities, and about good and evil. The stories of Norse mythology are of a great help with this, and the children are eager to learn all about it.

Local geography helps the students to orientate themselves in space. The students are introduced to grammar, singing in rounds, and musical notation. And in math, one of the most difficult concepts is introduced – fractions. The beautiful world of *one* falls to pieces, while they discover that there exists a new world of numbers in between any two whole numbers.

## *Curriculum Summary for Fourth Grade Math*

The world of numbers

- Factors. Students should be comfortable with finding all the factors of any number less than 50, and many of the factors for larger numbers, as well. This can then naturally progress to *common factors* (between two numbers), and the *greatest common factor* (GCF).
- LCM. *Least common multiples* should be introduced and practiced. This is needed as preparation for finding common denominators with fractions.

The arithmetic facts

- *Keeping it fresh.* If all went well in third grade, then the whole class should be quite solid with their arithmetic facts. In fourth grade, we simply need to keep it fresh. Our *fourth grade arithmetic facts review sheets* should be used three times per week (depending upon the class). Oral mental arithmetic also helps to keep the arithmetic facts fresh. (See *All about the Fourth Grade Arithmetic Facts Review Sheets* under *More Ideas for Teaching Fourth Grade Math* and *A Step-by-Step Progression for the Arithmetic Facts* found in the appendix.)
- *Times/Division tables.* In fourth grade, it is hopefully no longer necessary for the class to practice the 2-12's times/division tables in order (e.g., 7, 14, 21, 28, 35, etc.).
- *Challenging multiplication facts.* The teacher can select some challenging multiplication facts from the 13's through the 25's tables. Some of these can be done by the whole class, and some can be given just to those students needing an extra challenge. This work can be continued into fifth and sixth grade.

The four processes

- *Horizontal addition and subtraction.* It is still good for the students to practice strategies (perhaps learned in earlier grades) for addition and subtraction with numbers up to 1000. They should become fluent working problems like: $125+126$, $895+112$, $974-875$. This can either be done through written practice sheets or with oral mental arithmetic.
- *Vertical addition and subtraction.* This was introduced in third grade, but now, in fourth grade, we work with larger numbers. Regular practice is needed.
- *Vertical multiplication.* In third grade, we worked with single-digit multipliers. Now we can practice with two-digit and three-digit multipliers. Regular practice (but not too much!) is needed.
- *Long division.* Fourth grade is the time to introduce "flexible long division". Keep the divisors between 2 and 12 (e.g., $1788 \div 6$). Long division will be practiced a good amount in fifth and sixth grade.

Measurement. We should review third grade measurement, and begin doing simple conversion problems.

Fractions

- The goal in fourth grade is for the children to develop a sense for fractions – to bring them to an understanding of what a fraction is. *Too much of an emphasis on procedural skills with fractions will likely overwhelm many of the students, make it harder for them to develop a sense for fractions, and leave them hating fractions.*

## Fourth Grade Math

- *By the end of fourth grade*, the students should have a basic understanding of fractions, including problems similar to those found in the following list. Again, *keep it simple!*
  - Give two other fractions that are equivalent to two-thirds.
  - What is one-fifth of 35? What is ⅔ of 12?
  - Mr. Jones has twenty students in his class. If one-quarter of the class is outside, then how many students are outside?
  - How can two apples be divided fairly between four people? Between six people?
  - The students should become comfortable with fraction problems like these:

  $$\frac{3}{5} + \frac{1}{5} = \underline{\quad} \qquad \frac{7}{12} - \frac{3}{12} = \underline{\quad} \qquad \frac{3}{5} \times \frac{1}{2} = \underline{\quad}$$

  $$\frac{5}{6} + \frac{1}{2} = \underline{\quad} \qquad \frac{3}{4} - \frac{1}{3} = \underline{\quad} \qquad \frac{2}{3} \times 12 = \underline{\quad}$$

- *Types of fractions.* Although proper (i.e., "normal") fractions receive the most attention, improper fractions and mixed numbers should be briefly introduced.
- *Equivalent fractions.* The students should clearly understand the idea of equivalent fractions (e.g., that ¾ is the same as ⁶/₈).
- *Common denominators* are introduced, and by the end of the year, the students should be comfortable with finding a common denominator for two fractions.
- In order to avoid unhelpful confusion for the children, it is best to write the fraction bar horizontally rather than diagonally. For example, if we write ⁵/₆ – ¹/₃, it looks as if 1 should be subtracted from 6. We should instead write it as $\frac{5}{6} - \frac{1}{3}$. However, in this book (written for adults) we often write it as ⁵/₆ – ¹/₃ for formatting purposes and saving space.
- *From the whole to the part, and then from the part to the whole.* With fractions, we start with the whole and then divide it (e.g., an apple, cake, pizza, etc.) into parts. We then name each part. Then we bring the parts back into the whole by showing, for example, that ⅓ of an apple plus ⅓ of an apple plus ⅓ of an apple equals the whole apple. And then we can write ⅓ + ⅓ + ⅓ = 1.
- *The four processes with fractions.* Here in fourth grade, we introduce adding, subtracting, and multiplying fractions. Division of fractions should wait until fifth grade.

## *Recommended Reading for Fourth Grade Math*

- Henning Anderson, *Active Arithmetic*. AWSNA Publications.
- Herman von Baravalle, *The Teaching of Arithmetic and the Waldorf School Plan*. Waldorf School Monographs, 1967, third edition.
- Dorothy Harrer, *Math Lessons for Elementary Grades*. AWSNA Publications.
- Christoph Jaffke, *Rhythms, Rhymes, Games and Songs for the Lower School*. Published by the Pedagogische Forschung Stelle, Der Bund der Freien Waldorf Schule in Stuttgart.
- Ron Jarman, *Teaching Mathematics in Rudolf Steiner Schools for Classes I-VIII*, Hawthorn press, Gloucestershire, 1998. (We recommend this book for ideas for fourth grade and up.)
- Lipping Ma, *Knowing and Teaching Elementary Mathematics*. Lawrence Erlbaum Assoc. Publishers 1919, London.
- Ernst Schuberth, *Teaching Mathematics for First and Second Grades in Waldorf Schools*. Rudolf Steiner College Press. (Especially the section about math weaknesses, pages 59 and up.)
- John A. van de Walle, *Elementary and Middle School Mathematics*. Virginia Commonwealth University, Longman, NY.

# Lessons and Topics for Fourth Grade Math

<u>Scheduling</u>
The curriculum calls again for twelve weeks of math (three main lesson blocks of four weeks each), with one week of review at the end of the school year.

## Don't forget!!

- *Every day*, we should <u>review</u> the previous day's lesson.
- *Every day*, we should bring <u>something new</u> to the children.
- *Every day*, the students should <u>practice</u> the new material and selected review topics.
- *Every day*, we should ask the students to <u>estimate</u> something: "What do you think the answer will be?"
- *Every day*, we should ask this question at least once: "How did you do that?"

## Practice and review (of old material)

- During each math main lesson, we should be sure to practice the material from previous blocks and previous grades *for at least a half hour every day*.
- In order for the children to learn a topic well, it needs to be systematically reviewed. Obviously, all previous topics cannot be reviewed every day. The teacher needs to decide which topics are the most important to review each day. Some of these topics may need to be reviewed and practiced nearly every day, whether the class is in a math block, or not. As a general guideline, the class should practice 30 minutes daily, when in a math block, and about 10 minutes daily (during main lesson) when not in a math block. (Also see *Review and Practice* under *Principles of Waldorf Education* in the *Introduction*.)

<u>Regarding fourth grade fractions</u>
- *Caution!!* Often in Waldorf circles, you hear it said regarding the math curriculum, that "fourth grade is fractions". It is true that the introduction to fractions is an important part of the fourth grade year. However, in terms of classroom time spent in math lessons, work with fractions should not dominate everything else. Review and practice of concepts introduced in third grade, such as measurement and vertical arithmetic, is very important, as well as continued practice with horizontal arithmetic.
- *Keep it simple!* Everything listed here (for fourth grade) is intended as part of an introduction to fractions. There will be much more work with fractions in the coming years.
- It is not yet the goal to develop mastery of fraction skills – much of this is done in fifth and sixth grade. *Too much of an emphasis on procedural skills with fractions will likely overwhelm many of the students, make it harder for them to develop a sense for fractions, and leave them hating fractions.*
- The goal in fourth grade is for the children to develop a sense for fractions – to bring them to an understanding of what a fraction is. This is the foundation for studying fractions in future years.

<u>The world of numbers</u>
- *Factors.* Working with the concept of the factors of a number is an important theme of fourth grade, but it can be playful! Here are some ideas:
  - As a mental math exercise, ask the students to write down all the factors of a number. For example, if you say 21, then the student writes down 1, 3, 7, 21. Don't forget to occasionally throw in a prime number (for which the factors are only 1 and itself)!
  - With the students divided into small groups, they are to write down the factors of each number on a list. The numbers might start off fairly simple, but then get more challenging (e.g., 12, 14, 30, 26, 36, 42, 80, 180). And, it's fine to have some groups get further through these problems than other groups.
  - *Factor friends.* The students should discover for themselves that every factor has a "friend", and these two friends always multiply together to equal the number itself. For example, to find all the factors of 80, students may quickly identify 2, 4, and 5 as factors. But then we can ask, what are the factor friends of each of these numbers? In this way, we can find the factors 40, 20, and 16.
  - *Common factors.* For example, we can ask, "What are the common factors of 30 and 24?" We can then make one list of the factors of 24 (which is 1, 2, 3, 4, 6, 8, 12, 24), and another list of the

factors of 30 (which is 1, 2, 3, 5, 6, 10, 15, 30). Now we can see that the common factors are: 1, 2, 3, 6, and that 6 is the *Greatest Common Factor*.
- *Greatest Common Factor (GCF)*. After some practice, they should be able to do simple GCF problems in their heads. For example, what is the GCF of 24 and 32? Answer: **8**
- *Least common multiple* (LCM).
  - For example, we can ask, "What is the LCM of 4 and 6?" They then make a list of the multiples of 4 (i.e., numbers that are in the 4's table), and another list of the multiples of 6. They can then see that 12 is the LCM. This is preparation for finding the common denominator of two fractions.
  - We may also wish to revisit the movement exercise *two tables inside circles* that we did in second grade. This gives a more experiential understanding of the LCM and common denominators.
  - After some practice, they should be able to do simple LCM problems in their heads.
- Number dictations should still happen occasionally in fourth grade. The children should be fluent with writing and reading numbers well into the millions. We can include fractions in our number dictations.

Arithmetic facts
- Even if all went well in third grade with learning the arithmetic facts, there is still much work to be done in fourth grade so that the facts are not forgotten. They should be practiced briefly, nearly every day, either through oral mental arithmetic, or with written practice sheets. Our *Fourth Grade Arithmetic Facts Review Sheets* are available for free download at www.JamieYorkPress.com.
- *Division with remainders*. This was begun in third grade. For example, we can give 33÷6, and their answer should be "5 with a remainder of 3". We could also tell the students to take some number (e.g., 30) and divide it by 1, 2, 3, 4, etc. They can discover patterns with this.
- *Challenging multiplication facts*. The teacher can select some challenging multiplication facts from the 13's through the 25's tables. Some of these can be done by the whole class, and some can be given just to those students needing an extra challenge. This work can be continued into fifth and sixth grade.
- The arithmetic facts can now be practiced in a more challenging way. For example, we can toss a beanbag or a tennis ball to a student as we call out a number. The student then answers with a multiplication or division fact while she throws the ball back. For example, if the teacher says 28, then the student could give a variety of answers, such as: 2x14, 14x2, 7x4, 28x1, or 56÷2, 112÷4 etc. After working with one number, the students can write down all the different possibilities for that number.

Working horizontally. We should practice strategies (perhaps learned in earlier grades) for addition and subtraction, with numbers up to 1000. The students should become fluent with problems like 125 + 126, 895 + 112, 974 − 875. This can either be done through written practice sheets or with oral mental arithmetic.

Working vertically
- *Everyday,* we should practice vertical addition, subtraction, multiplication, and division. However, keep in mind, that some students will slowly, over time, gain mastery over vertical arithmetic. Be sure that these students don't become anxious about it; they should still be learning joyfully!
- *Vertical addition and subtraction.* This is continued from what was done in third grade, but now with larger numbers. Regular practice is needed.
- *Vertical multiplication*
  - In third grade, we worked with single-digit multipliers. Now in fourth grade, we can begin with two-digit multipliers, and build up to three-digit multipliers. Regular practice is needed.
  - Be sure to read *A step-by-step progression for vertical multiplication* under *More Ideas for Teaching Fourth Grade Math,* below.
- *Long division*
  - Long division should be new to the fourth grade class. We can build up to four-step problems, but keep the divisors between 2 and 12 (e.g., 15288÷6). Two-digit divisors (e.g., 1909÷23) should wait until fifth grade.
  - It can be helpful to introduce long division with a story, but be careful that the stories are not too elaborate, and that the details are relevant to the mathematical concepts.

# Fourth Grade Math

- *Flexible long division.* Long division doesn't have to be a blind procedure; it can be introduced so that the children understand how it works. It should be joyful! Be sure to read *A step-by-step progression for long division* under *More Ideas for Teaching Fourth Grade Math*, below.
- *Demonstrate that math is flexible!* Students should feel that there are different ways to solve a given math problem, even though showing multiple methods may not always be practical. In the case of long division, it is good to show different methods to the students. Parents, who are from different countries, can show the class how they learned to do division. The students don't need to understand these different methods, but they can appreciate that other methods exist.
- Once again, be sure that the class doesn't get overwhelmed with too many long division problems that are too difficult. It will be practiced much more in fifth and sixth grade.
- Check answers to division problems by multiplying (e.g., for 2292 ÷ 6 = 382, multiply 6 x 382 to make sure that it's equal to 2292).

Mental arithmetic
- Practice mental arithmetic daily (for 10 minutes or less), and continue to develop strategies.
- Work up to calculations like:
  400 – 12; 798 + 5; 3 x 400; 28 x 100; 1000 – 25; 3000 – 205; 8000 – 2700; 40 x 70; 300 x 5.
- *"Halfway" problems.* Simple variations of these problems may have been introduced in second grade and third grade, but now we can include more complicated variations, like:
    - What number is halfway between 14 and 32?
    - What number is halfway between 16 and 86?
    - What number is halfway between 250 and 350?
    - What number is halfway between 250 and 280?
    - What number is halfway between 600 and 1200?
    - What number is halfway between 600 and 1100?
    - What number is halfway between 25 and 30?
- In fourth grade, it can be helpful (and fun!) to ask the students to find as many strategies as possible. Here are a few possible strategies using the example "What's halfway between 14 and 32?":
    - *Estimate and check.* Estimate the answer, check it, and adjust until you finally get it. (This is probably what most students try to do at first.) With the given example ("What's halfway between 14 and 32?") a student might first estimate 22, but then checks the answer and finds that 22 is 8 away from 14, but 22 is 10 away from 32. So, she moves her answer up by one to get 23, and then checks to see it is correct – 23 is 9 away from both 14 and 32.
    - Add half the difference to the smaller number (probably the most common real strategy). The difference between 32 and 14 is 18. The answer is then 9 more than 14, which is 23.
    - Cut each number in half, and add the results together. The answer is then 7+16, which is 23.
    - Start at each number and count toward the middle simultaneously.
    - Start at some easy number in between, and then determine the average of the differences from that number and the two given numbers. Using the above example, we could choose 20, so the differences would be 6 and 12. Then find the average of these differences (which would be 9) and add that to the smaller number: 14 + 9 = 23.
    - Add the two numbers and divide by two, which is the typical averaging formula. $\frac{14+32}{2} = 23$.

Games. Try to find engaging games, including dice and board games (see the appendix for a list of games for lower school math). These can be wonderful activities for the children to do when there is extra time. (Under third grade "Games", see *Summing Cards*, which is still a great activity for fourth grade.)

Measurement. Review the material from the third grade measurement main lesson. Start with simple conversion problems (e.g., 3 pounds = ___ ounces), and then get more complicated (including some problems that have fractional answers), such as: 50 feet = ___ yards; 5 gallons = ___ cups, etc.

# Fourth Grade Math

Word Problems
- Simple written word problems are now introduced. Here are some examples of word problems to be done in fourth grade:
  - If we cut 8 apples into quarters, how many pieces do we have?
  - If Mary's mother buys oranges for $4.50 and peaches for $3.10, how much does she have to pay? If she gives the cashier a $10 bill, what should she get back for change? (Give many problems like this where the students need to determine how much change is given.)
- *Our puzzle and game book.* Fourth grade can also be a great time to bring riddles and puzzle problems to the class. Our *Making Math Meaningful Puzzle and Game Book* is intended as a resource for math teachers in grades four through twelve, in part to supplement the normal classroom material. It provides ideas for that "something different".

---

## Math Main Lesson Block #1 – Fractions, Part I

New material and content
- *The introduction to fractions.* In this block, the students discover a new number world. We begin by demonstrating fractions in a different way each day. (See *Fractions with food*, below.) An apple is a great first experience, as it is tangible, easy to understand, and has less of a chance of becoming a "crutch" in the way that a circular thing (e.g., pizza) can.
- *Terminology.* Language is an important part of understanding fractions. They become familiar with new words like numerator, denominator, and equivalent.
- Our goal is to bring the students to a basic understanding of what a fraction is (e.g., what $5/8$ really means).
- *Equivalent fractions.* The students should clearly understand the idea of equivalent fractions (e.g., that ¾ is the same as $6/8$).
- *Vertical multiplication*
  - We now progress from single-digit multipliers (third grade) to two-digit and three-digit multipliers.
  - Be sure to read *A step-by-step progression for vertical multiplication,* below.

Review and practice! Previously introduced material needs to be reviewed, practiced, and furthered. (See above, *Practice and review of old material* under *Lessons and Topics for Fourth Grade Math.*)

Daily progression for fractions
- *The first week.*
  - Day #1.
    On the first day, we introduce fractions with food (that will be eaten!) that is divided into pieces. We say clearly that you divide it first in half, then half again, and so on. That's it for the first day!
  - Day #2.
    On the second day of the block, we can "move" the fraction from the first day. For example, if we divided a cake into 12 pieces on the first day, then we can "walk it" on the second day. Every step represents a piece of the cake. The class then walks and speaks the fraction:

    $\frac{1}{12}, \frac{2}{12}, \frac{3}{12}, \ldots \ldots$ up to $\frac{11}{12}$, and $\frac{12}{12}$ (but we say "one")

    And then we do the whole thing backwards! While doing this, we make the students aware of how to properly pronounce the fractions. After this has been worked on, we can then do a new example of dividing something into pieces (e.g., dividing four apples – each apple into six pieces).

- *Day #3.*
  If we divided four apples into six pieces each on the previous day, we can now walk by sixths until we reach 4.

  $\frac{1}{6}, \frac{2}{6}, \frac{3}{6}, \frac{4}{6}, \frac{5}{6},$ "One",

  $1\frac{1}{6}, 1\frac{2}{6}, 1\frac{3}{6}, 1\frac{4}{6}, 1\frac{5}{6},$ "Two",

  $2\frac{1}{6}, 2\frac{2}{6}, 2\frac{3}{6}, 2\frac{4}{6}, 2\frac{5}{6},$ "Three",

  $3\frac{1}{6}, 3\frac{2}{6}, 3\frac{3}{6}, 3\frac{4}{6}, 3\frac{5}{6},$ "Four!".

  And then we do it backwards! After this has been worked on, we can then do another different example of dividing something into pieces.

- *Day #4 and #5.*
  We build very systematically on what we have done in the first three days.

- *During the second week*, we give the class a fraction and they walk forward (or backwards) to a certain number.

  Example: We tell the class to start at $3\frac{3}{4}$ and step up to 6. Their steps are then:

  $3\frac{3}{4}, 4, 4\frac{1}{4}, 4\frac{2}{4}, 4\frac{3}{4}, 5, 5\frac{1}{4}, 5\frac{2}{4}, 5\frac{3}{4}, 6.$

  Example: We tell the class to start at $3\frac{6}{8}$ and step up to 6. Their steps are then:

  $3\frac{6}{8}, 3\frac{7}{8}, 4, 4\frac{1}{8}, 4\frac{2}{8}, 4\frac{3}{8}, 4\frac{4}{8}, 4\frac{5}{8}, 4\frac{6}{8}, 4\frac{7}{8}, 5, 5\frac{1}{8}, 5\frac{2}{8}, 5\frac{3}{8}, 5\frac{4}{8}, 5\frac{5}{8}, 5\frac{6}{8}, 5\frac{7}{8}, 6.$

  Example: We tell the class to start at $1\frac{2}{6}$ and step down to zero. Their steps are then:

  $1\frac{2}{6}, 1\frac{1}{6}, 1, \frac{5}{6}, \frac{4}{6}, \frac{3}{6}, \frac{2}{6}, \frac{1}{6}, 0.$

- The students should write down what they have practiced into their practice books (e.g., they should write the same row of fraction steps from memory on the same day that they did it).

## Working with manipulatives

- *Fractions with food.* For the first week, we suggest starting the block with various edible objects (e.g., apples, pies, pizza, pancakes, chocolate) – preferably, each day, something of a different shape than the day before. For example, one day we work with apples (sphere), the next day a slice of bread (rectangle), the next day pizza (circle), the next day string cheese (line), and on the last day, a tall, round cake (which can be cut in many different ways).

- *Fractions with paper.* It is best for the children to create their own manipulatives for fractions. After we have tasted our fractions, we can then cut fractions from various shapes with paper: circles, triangles, rectangles, and squares. Again, the students need to experience different shapes for the same fraction. These paper fractions should be glued into their main lesson books.

- *Fraction envelopes* are a nice way for the children to have a hands-on experience of equivalent fractions. It also involves folding paper – a valuable exercise in itself.
  - Each child has their own fraction envelope, which contains an assortment of pie-shaped, colored pieces of paper.
  - In order to make the fraction pieces that go inside the envelope, each student will need seven paper circles – each one a different color, made from fairly firm paper, and 12cm in diameter. Each circle is folded and cut into equal-sized pizza-shaped pieces that represent ½, ⅓, ¼, ⅙, ⅛, ¹/₁₂. Each piece is then labeled with the fraction it represents, as shown above. Therefore, there should be eight same-colored, equal-sized pieces marked as ⅛. The last circle, which represents "the whole",

remains uncut.  Lastly, all the pieces are then put into an envelope, which the students keep in their desks.
- The students can then play with these fraction envelope pieces in a variety of ways – for example, to come to an understanding of equivalent fractions.
- Note that when the students use their fraction envelopes, they should always put the whole circle on the desk, as a reference.
- Mercurius sells round heavy folding paper (12cm diameter) in a set of 10 assorted colors, which works nicely for this project.  Etsy sells a nice set of wooden fraction circles, which may be nice to have in the classroom.

Bookwork
- In their main lesson books, the students should draw some of the "food fractions" that were done in class.  One page should have different cuttings (e.g., paper fractions) for each fraction.  It should also include the key concepts, such as: numerator, denominator, and equivalent.
- Vertical multiplication (with two and three-digit multipliers) needs to be thoroughly explained in the main lesson books, showing the step-by-step processes.
- Be sure to visit our website (www.JamieYorkPress.com).  Under the "Resources" tab you can see full-color pages from students' main lesson books.

## Lesson Plan Outline for Block #1 – Fractions, Part I
(Does not include all topics needing review from previous blocks and years)

To be done at least occasionally all year:
- Measurement review and practice.
- Number dictations.
- Simple word problems.
- Puzzles and games (see our puzzle/game book).
- Mental math or an *arithmetic facts review* sheet (one or the other each day).
- Vertical arithmetic (+/-/x) a bit every day.
- Simple division with remainders (29÷4 = 7 r 1).

Week #1
- Intro to fractions.  Every day we experience fractions in a different way (see *Daily progression for fractions*, above).
- Factors – e.g., find the factors of 20.
- Common factors between two numbers.
- Review and practice vertical multiplication from third grade (with one-digit multipliers).

Week #2
- More experiences of fractions.
- Walking fractions (see *Daily progression for fractions*, above).
- More work with factors and common factors.
- Vertical multiplication: Introduce two-digit multipliers (just step #1, **).

Week #3
- Introduce equivalent fractions.
- Make fraction envelopes.
- Introduce Greatest Common Factor (GCF).
- Find common multiples between two numbers.
- Vertical multiplication: two-digit multipliers (do step #2, **).

Week #4
- Practice equivalent fractions.
- Introduce and practice adding fractions with like denominators.
- Practice Greatest Common Factor (GCF).
- Introduce Least Common Multiple.
- Vertical multiplication: two-digit multipliers (do steps #3 and #4, **).

** See below: *A step-by-step progression for vertical multiplication with two-digit multipliers*, under *More Ideas for Teaching Fourth Grade Math*.

Fourth Grade Math

# Math Main Lesson Block #2 – Fractions, Part II and Long Division

<u>New material and content</u>
- *Adding and subtraction fractions with unlike denominators.*
    - Practice a fair amount of *Least Common Multiples* (LCM) before the start of this main lesson block.
    - Don't overdo practicing adding and subtracting fractions with unlike denominators – this is intended as an introduction. There will be plenty of time to practice this important skill in the future.
    - We present below two different methods for the introduction to adding fractions with unlike denominators. The first method is more hands-on (using the students' fraction envelopes), and the second method is free of any manipulatives.
    - For more visual ideas, see John E. van de Walle's book, *Elementary and Middle School Math*.
- *Finding common denominators by using the fraction envelopes*
    - This is a hands-on way of experiencing adding fractions. We should start out with very simple problems, and then slowly get more complicated. A possible progression could be:
    ½ + ¼;   ½ + ⅓;   ⅓ + ⅙;   ⅓ – ⅙;   ⁵⁄₁₂ + ⅙;   ⅚ + ½;   ¾ – ⅓
    - With each problem, the students should work with the pieces in their fraction envelope. For example, with the problem ⅚ + ½, they take 5 pieces of ⅙ and 1 piece of ½ and see how they fit together. They should discover that ½ = ³⁄₆. Therefore, ⅚ + ½ is the same as ⅚ + ³⁄₆. The students can now easily see that together they make ⁸⁄₆, which can also be written as $1\frac{2}{6}$ or $1\frac{1}{3}$.

- *Finding common denominators without using manipulatives.*
    - Step #1 Practice two things (that should have been introduced in the previous block):
    (1) Adding and subtracting fractions with like denominators, such as: $\frac{2}{5} + \frac{1}{5}$ and $\frac{7}{9} - \frac{2}{9}$.
    (2) Finding equivalent fractions. You give any fraction and the students come up with as many equivalent fractions that they can think of. Make this fun! For example, you write $\frac{2}{5}$ on the board, and then the students add to the board $\frac{4}{10}, \frac{10}{25}, \frac{6}{15}, \frac{20}{50}, \frac{2000}{5000}$, etc. At some point, the students realize that we can write it more systematically. Starting with $\frac{2}{5}$, we end up with the 2's table on the numerator, and the 5's table in the denominator: $\frac{2}{5}, \frac{4}{10}, \frac{6}{15}, \frac{8}{20}, \frac{10}{25}$, etc. Remember, the goal is to become comfortable both with finding equivalent fractions and adding and subtracting fractions with like denominators without needing any story, manipulative, or physical representation.
    - Step #2 (Discovering a method)
    Using the example $\frac{2}{3} + \frac{1}{4}$. Write $\frac{2}{3}$ on the left side of the blackboard, and underneath write many fractions equivalent to it: $\frac{2}{3}, \frac{4}{6}, \frac{6}{9}, \frac{8}{12}, \frac{10}{15}$, etc. Write $\frac{1}{4}$ on the right side of the blackboard, and underneath write many fractions equivalent to it: $\frac{1}{4}, \frac{2}{8}, \frac{3}{12}, \frac{4}{16}, \frac{5}{20}$, etc. Emphasize that all the fractions on the left side of the blackboard are equal to one another, and all the fractions on the right side of the blackboard are equal to one another. Our goal is to figure out a method for adding $\frac{2}{3} + \frac{1}{4}$. We know that whatever $\frac{2}{3} + \frac{1}{4}$ is equal to, it must be the same as any one fraction on the left plus any one fraction on the right. For example, $\frac{6}{9} + \frac{5}{20}$ must have the same answer as $\frac{2}{3} + \frac{1}{4}$. Now we can ask the students: "What fraction do we choose on the left and what fraction do we choose on the right so that we can easily add these two fractions?" They will see that the best choice is $\frac{8}{12} + \frac{3}{12}$, which we can easily add to become $\frac{11}{12}$. Therefore, we can say that $\frac{2}{3} + \frac{1}{4} = \frac{11}{12}$.
    - Step #3 (Use another example.) Review what we did the day before (step #2), then give another example, like $\frac{3}{5} + \frac{1}{3}$ using the same exact method as given above. That's enough for the day.

# Fourth Grade Math

- Step #4 (Groupwork)  Divide the class into 3-person groups.  Put several problems on the board (maybe 8 of them), starting out easy, and becoming increasingly difficult – the last one perhaps something like $\frac{6}{7} - \frac{3}{11}$.  Each group does as many of the problems as they can.  It's totally fine if some groups only do a few of the easier ones.  Our goal is that all students are actively engaged in the work, and everyone walks away feeling excited about fractions.

- Step #5 (Finding a shortcut)  This may simply arise naturally out of the class – some student may see a shortcut during our previous days' work.  If not, we can now ask the class: "Does anyone see a shortcut for adding fractions with unlike denominators?"  Maybe it's best to leave this as a hanging question to be answered at the start of block #3.  Either way, see if the answer can arise from the students, rather than giving it to them as a blind procedure.
  
  Here is one possible shortcut:
  
  *We can multiply the top and bottom of the left fraction by the right fraction's denominator, and multiply the top and bottom of the right fraction by the left fraction's denominator.*
  
  Example: $\frac{3}{4} - \frac{2}{5} \rightarrow \frac{3 \times 5}{4 \times 5} - \frac{2 \times 4}{5 \times 4} \rightarrow \frac{15}{20} - \frac{8}{20} \rightarrow \mathbf{\frac{7}{20}}$
  
  Or, maybe some of the students will see the relationship between finding the common denominator and the LCM (least common multiple) of the two denominators.  That would be exciting!

- *A systematic approach to finding the LCM or common denominator*
    Example:  What is the common denominator for ⅚ and ¼ ?
    This is the same as asking: "Where do the denominators meet?"  The students can then make a table, similar to what is shown on the right.

    | 6 | 4 |
    |---|---|
    | 6 | 4 |
    | **12** | 8 |
    |    | **12** |

- *Flexible long division*
  - If taught well, the first introduction to long division can be a joyful mathematical experience and be engaging for all students in the class – and the students can understand how it works!
  - We begin with a question: "How can we divide $465 evenly between 5 people?"  We then act it out with play money so that every student can follow what happens.  This then evolves into *flexible long division*.
  - Be sure to read *A step-by-step progression for long division* under *More Ideas for Teaching Fourth Grade Math*, below.  The whole progression can be done (in small steps) over a two-week period.
  - How much the students should practice (flexible) long division will depend on the class.  We can build up to problems as complicated as 32,832÷12 for the most advanced students.  For most of the class, practice problems should be kept fairly simple.
  - "Standard long division" (which is usually taught as a blind procedure and what most people are familiar with) waits until fifth grade, when the students will practice this skill much more.

Review and practice!  Previously introduced material needs to be reviewed, practiced, and furthered.  (See above, *Practice and review of old material* under *Lessons and Topics for Fourth Grade Math*.)

Activities and Movement
- In the first block, the class "walked" fractions.  Now we can walk different fractions at the same time.
  Example: (Using ⅛, ¼, ½) The students line up next to each other.  The ⅛ students take a small step for every count, while the ¼ students take a medium step (twice the size of the ⅛ students) with every other count, and the ½ students step (four times the size of the ⅛ students) with every four counts, starting at four-eighths.  And they continue to move along in this manner.
- To experience the concept of common denominators through movement, revisit the movement exercise, *two tables inside circles* (from second grade, block #3, above).

Bookwork
- The new concept of this block (adding and subtracting fractions with unlike denominators, and flexible long division) need to be thoroughly explained in the main lesson books. They should show the step-by-step processes. Use space, color, and drawings. Be creative!
- Be sure to visit our website (www.JamieYorkPress.com). Under the "Resources" tab you can see full-color pages from students' main lesson books.

## Lesson Plan Outline for Block #2 – Fractions, Part II and Long Division
(Does not include all topics needing review from previous blocks and years)

To be done at least occasionally all year:
- Measurement review and practice.
- Number dictations.
- Simple word problems.
- Puzzles and games (see our puzzle/game book).
- Mental math or an *arithmetic facts review* sheet (one or the other each day).
- Vertical arithmetic – a bit every day.
- Mental division with remainders (29÷4 = 7 r 1).

Week #1
- State our block goal: to add two fractions together that have unlike denominators.
- Practice adding/subtracting fractions with like denominators.
- Practice finding equivalent fractions.
- Introduction to reducing fractions.
- Intro to long division: do steps #1-4 (see **) leading up to *Story Long Division*.

Week #2
- Practice adding/subtracting fractions with like denominators.
- Practice finding equivalent fractions.
- Practice reducing fractions.
- The big step! Adding two fractions with unlike denominators. Each day, a different example.
- Intro to long division: do steps #5-8 (see **) (*Flexible Long Division*).

Week #3
- Practice finding equivalent fractions.
- Practice reducing fractions.
- Discover a shortcut for finding the common denominator to two fractions.
- Each day, a new example of adding or subtracting two fractions with unlike denominators. (Not much practice yet.)
- Practice flexible long division.

Week #4
- Wrap up the block.
- Review everything from weeks #1-3.

** See below: *Lesson Plan Outline for Introducing Long Division*, under *More Ideas for Teaching Fourth Grade Math*.

## Math Main Lesson Block #3 – Bringing it All Together!

New material and content
- *Multiplication with fractions.*
  - The typical approach for multiplying fractions is to simply multiply the numerators and multiply the denominators. However, without any buildup or explanation, this becomes a *blind trick*. Instead, we should lead the students to discover this for themselves. It thus becomes a *strategic trick*.
  - The whole process for teaching how to multiply fractions should not take very long – indeed much less time than the buildup to the introduction to adding fractions with unlike denominators.

# Fourth Grade Math

- Here is one step-by-step way that the students can be led to discover the *strategic trick* for multiplying fractions:
    - It is best to begin by translating the mathematical concepts into spoken language. We can start with this question: "We know that half of 6 is 3 (which we can write as ½ x 6 = 3). But what is half of a half?"
    - We can then demonstrate it visually, perhaps with an apple. Take half of an apple and cut that piece in half. Ask: "How much of the whole apple is this?" They will see that it is the same as ¼ of the apple. Then we say the verbal version: "Half of a half is equal to a quarter", and we write the equation version ½ x ½ = ¼ on the board. (The next day the students can write both of these versions into their main lesson books along with a drawing of the cut apple.)
    - Then we can do a similar apple demonstration and cut a quarter of an apple in half. So, we say: "Half of a quarter is equal to an eighth", and we write ½ x ¼ = ⅛
    - Now for the final step. We can write the following problems on the board and have the students (perhaps in small groups) try to answer each of the following questions:
        - ½ x 12; ¼ x 20; ⅓ x 18; ⅙ x 30
        - ½ x ¼ (remember the apple!); ½ x ⅓; ⅓ x ⅕; ¼ x ⅐; ⅑ x $^{1}/_{11}$
        - ⅕ x 35; ⅖ x 35; $^{1}/_{10}$ x 60; $^{3}/_{10}$ x 60; ⅛ x 24; ⅝ x 24;
        - ¼ x ⅕; ¾ x ⅕; ⅐ x ½; $^{5}/_{7}$ x ½; (now hopefully they see the trick!);
        - ⅔ x ⅘; ¾ x ⅗; $^{2}/_{7}$ x $^{3}/_{11}$
- And now we see the *strategic trick* for multiplying fractions: we simply need to multiply the two numerators together and multiply the two denominators together.
- It is likely that some of the students will be a bit lost with the above process. In this case, don't get bogged down trying to make sure that everyone gets all of the steps. (All the above steps can be done in one lesson!) The idea is to lead the class through a process where someone in the class sees what the shortcut is, and then the class can celebrate this discovery together. At the very least, the students see that the trick came from some process; it wasn't just handed to them blindly. Afterwards, everyone can use this newly discovered strategic trick.
- After it has been introduced, we can also offer a visual demonstration for multiplying fractions. Here's an example for ⅔ x ⅘, which we state orally as "two-thirds of four-fifths". We start with a rectangle which represents 1 (the whole). Then we divide the rectangle into fifths, and lightly shade-in four of the fifths to represent ⅘ of the whole. Then we divide the shaded region into thirds. At this point we ask the children: how much is each little shaded box of the whole? We can see that each little box is $^{1}/_{15}$ of the whole. Lastly, we darkly shade in ⅔ of the lightly shaded area. We can see that in the end, 8 of the little boxes have been darkly shaded, and since each one of the boxes represents $^{1}/_{15}$ of the whole, we have demonstrated that ⅔ x ⅘ = $^{8}/_{15}$.

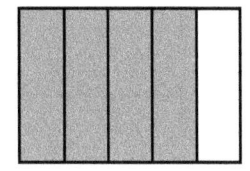

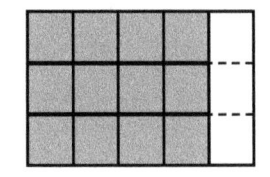

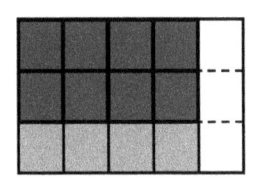

- The shortcut called "cross-canceling" should wait until fifth or sixth grade.
- Division of fractions should wait until fifth grade.
- *Mixed numbers and improper fractions.*
    - A mixed number (e.g., 4⅔) has two parts – a whole part and a fractional part. An improper fraction (e.g., ⁵/₄) is a fraction where the numerator is larger than the denominator.
    - We should introduce improper fractions and mixed numbers by working with manipulatives, such as an apple, or a loaf of bread. We can then demonstrate that, for example, if we have 3 whole loaves of bread and a quarter loaf, then that is the same as 13 quarter loaves. Then we can write 3¼ = ¹³/₄

- The most complicated problems we give with this topic involve converting mixed numbers into improper fractions. Of course, it is best if we can lead the students to discovering the trick for doing this, which is to multiply the whole number times the denominator, and then add the numerator – that gives us the new numerator. For example, with 5¾, we do 5x4+3, which is 23, so we get ²³/₄.
- Going in the other direction – converting an improper fraction into a mixed number simply becomes a division problem with a remainder. For example, to convert ¹⁷/₅, we ask ourselves: "How many wholes do I get out of 17 fifths?" The answer is that you get 3 wholes with 2 fifths left over, which we write as 3 ²/₅. This leads us to an important discovery:
  **The fraction bar really tells us to divide!**
- Doing arithmetic (+,-,x,÷) with improper fractions and mixed numbers should wait until fifth grade.

Review and practice! Previously introduced material needs to be reviewed, practiced, and furthered. (See above, *Practice and review of old material* under *Lessons and Topics for Fourth Grade Math*.)

Activities and Movement. The movement exercises for fractions slowly transform from the physical to lively imaginative thinking. Starting now, the students have to do the math movement internally, instead of doing it physically. This progresses, until in middle school, it has become a lively, inner movement.

Bookwork
- The students should draw the new concepts of fraction multiplication into their main lesson books.
- Be sure to visit our website (www.JamieYorkPress.com). Under the "Resources" tab you can see full-color pages from students' main lesson books.

## Lesson Plan Outline for Block #3 – Bringing it All Together!
(Does not include all topics needing review from previous blocks and years)

Notes:
- This block provides for the opportunity to consolidate everything from the first two blocks.
- If any topic (listed under the first two blocks) wasn't covered, it can be done now.

Topics to be practiced throughout the block
   (But be careful not to overdo it!):
- Finding equivalent fractions.
- Reducing fractions.
- Finding GCF's and LCM's.
- Adding and subtracting fractions with unlike denominators.
- Vertical addition and subtraction.
- Vertical multiplication with two and three-digit multipliers.
- Flexible long division.

Other topics to be considered:
- Simple word problems.
- Puzzles and games (see our puzzle/game book). Consider giving a new puzzle every day!
- Mental math or an *arithmetic facts review* sheet (one or the other each day).
- Challenging multiplication facts (e.g., 6x15).

Week #1
- To set the stage for fraction multiplication, give problems like these:
  - ⅓ x 12 → ⅓ of 12 → 12÷3 = 4
  - ⅔ x 12 → ⅔ of 12 → twice ⅓ of 12 = 8
  - ⅕ x 35 → ⅕ of 35 → 35÷5 = 7
  - ²/₅ x 35 → ²/₅ of 35 → twice ⅕ of 35 = 14
  - ³/₅ x 35 → ³/₅ of 35 → 3 times ⅕ of 35 = 21
  - ⁴/₅ x 35 → ⁴/₅ of 35 → 4 times ⅕ of 35 = 28

Week #2
- Lead the students to discover the *strategic trick* for multiplying fractions. (See above)

Week #3
- Practice multiplying fractions.
- Introduction to improper fractions and mixed numbers.

Week #4
- Wrap up the block.
- Practice converting between mixed and improper.
- Mention what we'll learn next year:
  - division of fractions.
  - arithmetic with mixed numbers.
  - decimal fractions.

Fourth Grade Math

## *More Ideas for Teaching Fourth Grade Math*
(Be sure to reread relevant sections from the *Introduction*.)

### All about the fourth grade arithmetic facts *review* sheets, and daily practice

- Free download!! You can download our fourth grade arithmetic facts review sheets *for free* from our website: www.JamieYorkPress.com.
- This is review! These math fact sheets are intended as review of the *Third Grade Arithmetic Facts Practice Sheets*. It is important that the students first do those third grade *Practice Sheets* (and are very familiar with the 105 facts in them) before doing these fourth grade *Review Sheets*.
- Timing. Each sheet has 30 problems. There are 100 sheets in the whole set. A class should do three sheets per week (on average) throughout the fourth grade year.
- Caution! This should be fun and easy for the students. If successful, this builds their confidence. It is important to make sure that these sheets don't become torture for the students. Try to de-emphasize the importance of speed (this can wait until fifth grade). Help the students realize that improvement is important.
- A special note regarding division with remainders. Towards the end of the *Third Grade Arithmetic Facts Practice Sheets*, we introduced division with remainders. On those sheets, these remainder problems looked something like $31 \div 5 = \_\_ r \_\_$, which warned the students that they were remainder problems. In contrast, the *Fourth Grade Review Sheets* (found here) don't warn the students of the remainder problems. Therefore, the problem is simply given as $31 \div 5 = \_\_$, and the student should write the answer as 6 r 1. The students should know that on every sheet (of 30 problems) there is always one remainder problem. This makes it consistent, and perhaps fun for the students to find the mystery remainder problem.
- What comes next? The next set of practice sheets is titled *Fifth Grade Arithmetic Facts Speed Sheets*. They are designed to increase the calculation speed of the students.
- The whole picture. The 30 problems listed on a particular sheet are only a part of a day's math practice, which should only take about five minutes of class time. **It would be very unfortunate if daily math practice consisted of nothing more than the 30 arithmetic facts practice problems that appear on these sheets.**
- The elements of math practice. The following list shows some of the aspects to consider when planning math practice for the day.
  - *Arithmetic facts review sheet.* The teacher copies by hand the 30 problems from a given day (i.e., these fourth grade arithmetic facts review sheets) onto paper to be photocopied.[32] (about 5 minutes)
  - *Mental arithmetic.* The teacher may decide to read the first 6 problems out loud.
  - *Extra math practice problems.* There should be a few problems that the teacher comes up with and writes on the board. The students copy them into their practice books and work out the answers. These problems also include practice and review of material covered in previous math blocks. (This takes 20-25 minutes if the class is in a math block, otherwise only 5 minutes.)
  - *Challenge problems.* It is important that the last few problems (that the teacher adds on) be more challenging in order to keep the "quicker" students fully engaged.
- And what happens if…? We hope that it won't happen, but even when most of the class is doing well with these sheets, there may be a few children who still haven't solidly learned their arithmetic facts. In order to help these children, it may be helpful to give them a multiplication/division table (i.e., a square for the tables). Additionally, these students could study the basic arithmetic problems with flashcards during morning practice time.

---

[32] *Saving paper.* Try to find ways to reduce the amount of paper being used. For example, rather than using one sheet of paper each day, each side could be divided into three columns (i.e., 6 days per sheet). This is one of many ways to help develop an environmental consciousness in the students.

# Fourth Grade Math

## A step-by-step progression for vertical multiplication (with two-digit multipliers)

**Notes:**
- The students should draw columns for the place values: thousands, hundreds, tens, and ones. It may be that graphing paper (with square grids) is helpful for students.
- We strongly recommend that students write a "0" (in all grades) instead of a blank for the placeholder. This better shows that we are actually multiplying with tens, hundreds, thousands, etc.
- We recommend that, in the end, the students avoid writing the small "carry" number on top of the multiplication problem. Instead, they should simply keep it in their heads.

**Step #1:**

```
                    4 7    →   40+7
                 x  3 6    →   30+6
6x7    =            4 2
6x40   =          2 4 0
30x7   =          2 1 0
30x40  =       + 1 2 0 0
                 1 6 9 2
```

**Step #2:**

```
                      2 4
                    4 7
                 x  3 6
47x6   =            2 8 2
47x30  =         + 1 4 1 0
                   1 6 9 2
```

**Step #3:**

```
              2 4
            4 7
         x  3 6
            2 8 2
         + 1 4 1 0
           1 6 9 2
```

**Step #4:** (Without the carry digits!)

```
            4 7
         x  3 6
            2 8 2
         + 1 4 1 0
           1 6 9 2
```

77

Fourth Grade Math

# Lesson Plan Outline for Introducing Long Division

- Step #1: Warm-up day. Practice doing a fair bit of mental math with multiplying problems with zeroes (e.g., 400x6 and 70x30). State the question at the end of class: In the next two weeks we will learn how to do long division with big problems, like 3185÷5. How can we do this?

- Step #2 (Acting it out): Act out 465÷5=93 with monopoly money and five students. The story is simple: I found $465 in a bag, and decided to divide it evenly between my five friends. Try giving away these amounts each time: 60, 20, 5, 3, 5. After each "round" of giving away, we ask how much do I now have in my bag? At the end, we ask, how much money does each person have? Nothing gets written down.

- Step #3 (**Story Long Division**): Review orally what we did yesterday, recalling all of the details. Then go through it all again, writing it in "Story Long Division" form with two columns: "Bag" and "Person". Then, do a second problem 3336÷4=834 (again, with money and with students at the front of the room acting as the friends). Then write this problem on the board in "Story Long Division" form.

| **Bag** | | **Person** |
|---|---|---|
| 3336 | | |
| − 2000 | ← x4 ← | 500 |
| 1336 | | |
| − 400 | ← x4 ← | 100 |
| 936 | | |
| − 800 | ← x4 ← | 200 |
| 136 | | |
| − 100 | ← x4 ← | 25 |
| 36 | | |
| − 20 | ← x4 ← | 5 |
| 16 | | |
| − 16 | ← x4 ← | + 4 |
| 0 | | 834 |

- Step #4 (Students solve problems in groups): Review the second problem 3336÷4=834 by rewriting it (in Story Long Division form) on the board. Then do it a second time, but giving away different amounts. (You don't need to use money or have students at the front of the room for the second method.) Show the second method on the board next to the first method. Then have the students do a new problem (564÷3=188) in four-person groups. Each group needs to have $564 in monopoly money. (The teacher is the bank, making change as needed.)

- Step #5 (**Flexible Long Division**): Write all the different ways on the board that the groups did yesterday's problem. Do a new problem (2292÷6) and the solution both in Story Long Division form and a new way: Flexible Long Division (see below). Then do new problems in groups, but this time without using money, such as: 348÷4=87    486÷3=162    1880÷5=376

$$2292 \div 6 = 382$$

```
6 | 2292        200
  − 1200        150
    1092         20
  −  900       + 12
     192        382
  −  120
      72
  −   72
       0
```

- Step #6: On the board, show 2024÷8=253 with both forms (Story Long Division and Flexible Long Division). Then have the students do these problems in groups: 261÷3=87    1215÷5=243; 1152÷6=192    28,638÷9=3182
Emphasize that it's OK if you make mistakes, and OK if it takes you a while. Mention that tomorrow we'll do one problem in groups, and then you'll do it on your own.

- Step #7: Only do Flexible Long Division from now on.
On the board: 1281÷3=427
In groups: 685÷5=137
On your own: 504÷4=126    3185÷5=637
                517÷11=47    32,832÷12=2736

- Step #8: More practice. At the end, mention that next year we will learn "Standard Long Division"

- Future Steps:
  - Do some practice of Flexible Long Division later in fourth grade.
  - In fifth grade, review Flexible Long Division, and then ask how can we do it more efficiently, where each step gives the "perfect" number of hundreds, tens and ones? E.g., with 1925÷5, the answer is then written 300+80+3 above the house, as a transition to "Standard Long Division".
  - In sixth grade, answers can include repeating decimals.
  - In sixth grade, introduce "Short Division".

# –Fifth Grade Math–

## *Overview of Child Development in Fifth Grade*

We leave the troubled lives of the Norse gods behind as the sun heralds a beautiful start of a new day. This is now the middle of the grade school years. In fifth grade, temperaments and talents become more visible, and the student communicates easily between the inner and the outer world. There is an increased interest in social interactions. Breathing and blood circulation become harmonized. The proportions of the physical body are in harmony, thereby allowing for graceful physical movements. The child's cognitive capacities are increasing, which is accompanied by a desire to be challenged. The students want to improve their work because they want to be proud of their work. They want and need constructive feedback.

The fifth grade history curriculum, which starts with the ancient cultures and culminates in Greek history, develops their orientation in time. In language arts, the students are introduced to the verb tenses. In music, the class can now learn to sing multi-part harmonies.

The students have matured and now have an ability for a deeper understanding of more complex, mathematical concepts. This is the time to become confident in math concepts from previous years (e.g., vertical (long) multiplication, fractions, etc.). We give the students more challenging problems, which slowly, with regular practice, help to increase their skill level. It is in fifth (and sixth) grade that the students solidify their foundation in basic arithmetic. The new topic of decimal fractions sets the stage for the introduction to percents in sixth grade.

## *Curriculum Summary for Fifth Grade Math*

Decimal fractions[33]. The number world of decimal fractions is introduced. Place value practice supports this.

Fractions
- The students should become fluent when working with common and unlike denominators.
- Division of fractions is introduced.
- All four processes (+, –, x, ÷) with common fractions and mixed numbers should be practiced regularly.

Measurement
- The metric system is introduced, which can be nicely integrated with decimal fractions.
- Distance, weight, and capacity (volume) of the metric system should all be introduced.
- Review of the U.S. measurement system should be integrated into our work with the metric system. (However, we should not yet give the students conversions problems that go between the two systems.)

Geometry
- *Freehand geometry*. The students should create many beautiful drawings of circles, squares, triangles, angles, divisions of the circle, etc., all freehand, without the aid of a ruler or compass. The accuracy required for this takes great focus and will.
- We might wish to bring an imaginative picture of the *Pythagorean Theorem* at this time.
- We can also introduce *perimeter and area*.

The Wonder of Number. This is the central theme for a creative main lesson. The idea is to bring several properties of numbers that will leave the students with a feeling of wonder. There are many topics and mathematical properties that can accomplish this. (See details under *Math Main Lesson Block #3: The Wonder of Number,* below.)

Review and Practice
- Consolidation of skills (that were introduced in earlier years) is a major theme for both fifth grade and sixth grade math. Systematic review and practice are needed throughout the year.
- It is especially important to review (and further) all work with vertical arithmetic (addition, subtraction, long multiplication, and long division), and all work with fractions.

---

[33] Throughout fifth grade, we should always say "decimal fractions" instead of just "decimals". When we only say "fraction", we mean the "common fraction" that we introduced in fourth grade. We should also speak these numbers in a way that shows they are fractions. For example, with 1.42, try to avoid saying "one point four two", but rather say, "one and forty-two hundredths". This all may be phased out in sixth grade.

# Fifth Grade Math

Arithmetic facts
- Review and practice of the arithmetic facts is still important. This can be done as part of daily oral mental arithmetic, and further reinforced a few times per week with our *Fifth Grade Arithmetic Facts Speed Sheets,* which can be downloaded from our website.
- Also, we can work with challenging multiplication facts (e.g., selected from 13-25 tables), which may have been started in fourth grade.

## *Recommended Reading for Fifth Grade Math*

- Henning Anderson, *Active Arithmetic.* AWSNA Publications.
- Herman von Baravalle, *The Teaching of Arithmetic and the Waldorf School Plan.* Waldorf School Monographs, 1967, third edition.
- Julia E. Diggins, *String, Straight Edge & Shadow.* Jamie York Press.
- Dorothy Harrer, *Math Lessons for Elementary Grades.* AWSNA Publications.
- Christoph Jaffke, *Rhythms, Rhymes, Games and Songs for the Lower School.* Published by the Pedagogische Forschung Stelle, Der Bund der Freien Waldorf Schule in Stuttgart.
- Ron Jarman, *Teaching Mathematics in Rudolf Steiner Schools for Classes I-VIII*, Hawthorn press, Gloucestershire, 1998. (We recommend this book for ideas for fourth grade and up.)
- Lipping Ma, *Knowing and Teaching Elementary Mathematics.* Lawrence Erlbaum Assoc. Publishers 1919, London.
- Ernst Schuberth, *Geometry Lessons in the Waldorf School, Volume 2.* AWSNA Publications. (Note: This is an excellent book, but, for the most part, what he has listed for fourth grade, we would recommend for fifth grade, and what he has listed for fifth grade, such as work with a compass and straight edge, we would recommend for sixth grade.)
- Malba Tahan, *The Man Who Counted,* W.W. Norton and Company, New York, 1993.
- John A. van de Walle, *Elementary and Middle School Mathematics.* Virginia Commonwealth University, Longman, NY.
- Arnold Wyss, Ernst Buhler and others, *Lebendiges Denken durch Geometrie*, Verlag Freies Geistesleben, Stuttgart, 1978. In German, but has many ideas for drawings for the Freehand Geometry block.

## *Lessons and Topics for Fifth Grade Math*

Scheduling
The curriculum calls again for twelve weeks of math (three main lesson blocks of four weeks each), with one week of review at the end of the school year.

### Don't forget!!

- *Every day*, we should review the previous day's lesson.
- *Every day*, we should bring something new to the children.
- *Every day*, the students should practice the new material and selected review topics.
- *Every day*, we should ask the students to estimate something: "What do you think the answer will be?"
- *Every day*, we should ask this question at least once: "How did you do that?"

### Practice and review (of old material)

- Rudolf Steiner recommended one practice hour (skills class) per week for math beginning in fifth grade.
- During each math main lesson, we should be sure to practice the material from previous blocks and previous grades *for at least a half hour every day.*
- In order for the children to learn a topic well, it needs to be systematically reviewed. Obviously, all previous topics cannot be reviewed every day. The teacher needs to decide which topics are the most important to review each day. Some of these topics may need to be reviewed and practiced nearly every day, whether the class is in a math block, or not. As a general guideline, the class should practice 30

## Fifth Grade Math

minutes daily, when in a math block, and about 10 minutes daily (during main lesson) when not in a math block. (Also see *Review and Practice* under *Principles of Waldorf Education* in the *Introduction*.)

<u>Working vertically with the four processes</u>. *Every day,* we should practice some problems using vertical addition, subtraction, multiplication, and division. We can also integrate work with decimals into this, as well as problems with extra zeroes at the end.

- *Vertical addition and subtraction.* Continued from what was done in fourth grade, but now with larger numbers. Regular practice is needed.
- *Vertical multiplication.* Simply continue and further what was done in fourth grade. Perhaps now we can even multiply two 4-digit numbers together. Regular practice is needed.
- *Standard Long Division.*
    - Last year, in fourth grade, we introduced long division through a process that enabled students to understand how it works. This culminated with *Flexible Long Division* (see Fourth Grade, *Introduction to Long Division*).
    - Now in fifth grade, we should review *Flexible Long Division*, and then make the transition to *Standard Long Division*, which is the way that long division is typically taught. *Standard Long Division* is really a shortcut. We have consciously chosen this progression (over the course of two years) because it begins with understanding and ends with an efficient shortcut process that can be used as a skill in the coming years.
    - The sequential steps for *Standard Long Division* are as follows (using the example 152÷4):
        1. First, we ask: "How many times does 4 go into 15?" (Answer: 3)
        2. Multiply 3x4 and then write it under the 15.
        3. Subtract to get a remainder of 3.
        4. Bring down the next digit.
        5. Repeat the above four steps until the problem is finished.

$$\begin{array}{r} 38 \\ 4\overline{\smash{)}152} \\ -12\phantom{0} \\ \hline 32 \\ -32 \\ \hline 0 \end{array}$$

    We should try to be very consistent and formulate the same questions during the process each time we do a long division problem.[34]
    - We should start out with single-digit divisors, and then build up to two-digit divisors (e.g., 1909÷23). When giving practice problems to the class, it may be helpful to have as many as eight or ten problems, which progress from easy to challenging (e.g., 63206÷65) and allow students to get as far as they can without feeling pressured to do all of them. In this way, all students are satisfied.
    - Be sure to (at least occasionally) check answers to division problems by multiplying (e.g., with 1909÷23=83, we should multiply 83x23 to make sure that it's equal to 1909).

<u>(Common) Fractions</u>
- Our work with (common) fractions was begun last year. All four processes with fractions and mixed numbers need to be thoroughly practiced. The students need to feel confident with fractions.
- *Addition and subtraction.* The students should become fluent with adding/subtracting fractions with like denominators and comfortable with adding/subtracting fractions with unlike denominators.
- *Multiplying and dividing.* Last year, in fourth grade, we introduced how to multiply (common) fractions, but division of fractions was not yet touched upon. This year, in fifth grade, we shall introduce how to divide fractions, and then we will practice multiplying and dividing fractions.

---

[34] Herman von Baravalle, *The Teaching of Arithmetic and the Waldorf School Plan.*

# Fifth Grade Math

<u>Estimating</u>. The students should learn to estimate before calculating a problem.

<u>Arithmetic facts</u>
- *Speed sheets*. Even if the class successfully learned all of their arithmetic facts in third and fourth grade, practicing these facts in fifth grade is still important. This can be done as part of daily oral mental arithmetic, and further reinforced a few times per week with our *Fifth Grade Arithmetic Facts Speed Sheets,* which can be downloaded from our website.
- *Challenging multiplication facts*. Previously, we had suggested that the fourth grade teacher could select some new multiplication facts from the 13's through the 25's tables. Some of these can be given to the whole class, and some can be given just to those students needing an extra challenge. This work may be continued in sixth grade.

<u>Mental arithmetic</u>
- Practice mental arithmetic daily (for 10 minutes or less).
- Work up to calculations like: $84-26$; $423-60$; $546+76$; $40 \times 600$; $5.378 \times 100$; $24,000 \div 600$.
- *"Halfway" problems*. We can continue doing problems that were done in earlier grades (e.g., "What is halfway between 26 and 56?"), but now we can add problems that work with fractions, like:
    - What number is halfway between $3/8$ and $7/8$? (Answer: $5/8$)
    - What number is halfway between $3/7$ and $4/7$? (Answer: $1/2$)
    - What number is halfway between 25.5 and 26.3? (Answer: 25.9)

<u>Word Problems</u>. Many of the word problems given in fifth grade should be practical, such as:
- If a pound of cheese is divided evenly between 8 people, then how much does each person get?
- 23 fifth graders are working on a garden project. Every one works 1½ hours. How many working-hours is that altogether?
- Simple *unit cost* problems, including:
    - If oranges cost 74¢ per pound, then how much do you need to pay for 5 pounds of oranges?
    - If 4 pounds of oranges cost $5.16, then what is the price per pound?
    - If 4 pounds of oranges cost $5.16, how much do 7 pounds of oranges cost?
    - If 4.3 pounds of oranges cost $5.16, then what is the price per pound?
    - If 4.3 pounds of oranges cost $5.16, then how much do 7 pounds of oranges cost?

<u>Puzzles</u>
- Fifth grade is a good time to give puzzles, but be careful to select a puzzle that is at the appropriate level. Puzzles that are too hard can be discouraging for many of the students. The right puzzle can bring joy into the classroom for all of the students.
- Our *Making Math Meaningful Puzzle and Game Book* is intended as a resource for math teachers in grades four through twelve, in part to supplement the normal classroom material. It provides ideas for that "something different".
- *Math Magic Trick*. Here is an example from our *Puzzle and Game Book*: Start with two numbers between 1 – 9. Add these two numbers together to get a third number. Add the second and third number together to get a fourth number. Add the third and fourth numbers together to get a fifth number. Continue this process until you have ten numbers. Add together all ten numbers. Divide by 11. This final answer will always be equal to the seventh number in your list. (Several years later, we can use algebra to show why this works.)

Fifth Grade Math

# Math Main Lesson Block #1: Freehand Geometry and Fifth Grade Fractions

New material and content

*Freehand geometry.*

- After form drawing in grades one through four, we step right into freehand geometry in fifth grade. These geometrical drawings require extra care because they are so ordered. The drawings demand will forces from the students as they learn to follow the geometrical laws of the forms.
- Freehand geometry also develops the feeling capacity for the ideal form itself. The connection with the archetypal form has to come from the feeling realm – not yet out of the intellectual realm.
- Most of the drawings with this topic involve circles, rectangles, angles, and triangles – all with a good amount of variation.
- Both beauty and exactness are important. The drawings should be done beautifully in color, and as precisely as possible. The students should draw the forms initially with lead pencil, and then certain aspects of the form can be accentuated by shading with color pencil. However, any artwork or shading should be simple, with the purpose of bringing out the most important characteristics of the form. Some of the drawings can be done on larger pieces of paper.
- The use of the geometric tools (compass and straightedge) should wait until the sixth grade during the *Geometric Drawing* main lesson.
- If you need more ideas for freehand geometry drawings, here are some resources:
  - See the appendix in this book for a sampling of (black-and-white) drawings.
  - Visit our website. Look under the "Resources" tab, and then the fifth grade "view sample main lesson book pages".
  - Ernst Schuberth's book, *Geometry Lessons in the Waldorf School, Volume 2.* However, note that we recommend freehand geometry in fifth grade, and geometric drawing (with compass and straightedge) for sixth grade, whereas Schuberth lists both of these for a year earlier.
  - The German book, *Lebendiges Denken durch Geometrie,* contains many beautiful drawings and will inspire the teacher with many ideas.
- Learning the terminology and vocabulary of geometry (e.g., radius, circumference, equilateral, pentagon, hexagon, etc.) can be part of the lesson – but not too much, as it will be furthered in future years.
- Angles and area and perimeter (of rectangles and squares) can now be introduced.
- *The Pythagorean Theorem.*
  - Since ancient Greek history and culture is a theme of fifth grade, we can tell Pythagoras's biography and introduce his theorem. (See Julia Diggins' book, page 92-105.)
  - By giving an artistic introduction to the Pythagorean Theorem in fifth grade, we are planting a seed for geometry in the upper grades. The children's first experience with it should be artistic and very hands-on. Connection with will and feeling is important. Be sure not to make it too intellectual.
- *Timing.* Every day, with this fifth grade main lesson, we should roughly divide the time equally between these two central topics: freehand geometry and fractions.

*Fifth grade fractions*

- *Review and practice.* Many aspects of (common) fractions were introduced last year. Now that we are in fifth grade and the material has slept for a while, we can practice our skills with fractions. There is much to be done, but remember our ultimate goal with these fraction skills – to be solid with all of it by the end of sixth grade, and to have a good relationship to the topic.
- *Cross-canceling.* Cross-canceling is a trick for multiplying fractions. It can wait until sixth grade.

# Fifth Grade Math

- *Division with fractions.*
  - Our goal, once again, is to lead the students to discover the *strategic trick* for themselves.
  - As with fraction multiplication, we should introduce fraction division by translating the mathematical concepts into spoken language. Here is a step-by-step progression:

    | | |
    |---|---|
    | $12 \div 6 = 2$ | How many times does 6 fit into 12? |
    | $12 \div 4 = 3$ | How many times does 4 fit into 12? |
    | $12 \div 3 = 4$ | How many times does 3 fit into 12? |
    | $12 \div 2 = 6$ | How many times does 2 fit into 12? |
    | $12 \div 1 = 12$ | How many times does 1 fit into 12? |
    | $12 \div \frac{1}{2} = 24$ | How many times does ½ fit into 12? |
    | $12 \div \frac{1}{3} = 36$ | How many times does ⅓ fit into 12? |
    | $\frac{1}{2} \div \frac{1}{4} = 2$ | How many times does ¼ fit into ½? |

  - In this way, the students should discover for themselves that the *strategic trick* is to multiply by the reciprocal of the second fraction. Even if this discovery is led by a few of the "quicker" students in the class, it is valuable for the whole class to see that it comes from some process.

- *Mixed numbers.*
  - A mixed number (e.g., 4⅔) has two parts – a whole part and a fractional part. An improper fraction (e.g., ⁵⁄₄) is a fraction where the numerator is larger than the denominator.
  - Although we briefly introduced the idea of a mixed number last year, we avoided doing any arithmetic (four processes) with mixed numbers. In fifth grade, performing the four processes with mixed numbers is one of the most complicated topics of the year. So, don't overdo it! There is plenty of time to work on this in sixth grade.
  - Here are a few examples of problems involving mixed numbers we can build up to over the course of fifth grade:

    Convert $5\frac{2}{7}$ to improper (Answer = $\frac{37}{7}$)  $\qquad$  $9\frac{1}{5} - 6\frac{4}{5}$ (Answer = $2\frac{2}{5}$)

    Convert $\frac{23}{5}$ to mixed (Answer = $4\frac{3}{5}$) $\qquad$ $7\frac{3}{8} + 2\frac{2}{5}$ (Answer = $9\frac{31}{40}$)

    $5\frac{7}{8} + 2\frac{3}{8}$ (Answer = $8\frac{2}{8}$ or $8\frac{1}{4}$) $\qquad$ 4½ x 1⅓ (Answer = 6)

    $8 - 2\frac{1}{3}$ (Answer = 5⅔) $\qquad$ 6⅔ x 2¾ (Answer = 18⅓)

Be careful that the above problems don't just become blind procedures. Some things to note:

- With the problem $9\frac{1}{5} - 6\frac{4}{5}$, the standard method is to convert the $9\frac{1}{5}$ into $8\frac{6}{5}$ and then subtract. But instead, we can just "think it through" and subtract the 6 first ($9\frac{1}{5} - 6 = 3\frac{1}{5}$) and then subtract the $\frac{4}{5}$. So we get $3\frac{1}{5} - \frac{4}{5}$, where we can imagine stepping backwards four times from $3\frac{1}{5}$ by one-fifth each step: $3\frac{1}{5}$, 3, $2\frac{4}{5}$, $2\frac{3}{5}$, **$2\frac{2}{5}$**.

- With 6⅔ x 2¾ we often hear that you must first convert the mixed numbers into improper fractions, so we get $\frac{20}{3}$ x $\frac{11}{4}$ (which eventually simplifies to 18⅓). The teacher should at least be aware that we also could have done the problem without converting to improper fractions. This can be done by treating it similarly to the vertical multiplication of two 2-digit numbers (e.g., 23x56), which requires us to do four multiplications. So, using this method, we get: 6⅔ x 2¾ → (⅔ x ¾) + (6 x ¾) + (2 x ⅔) + (6 x 2), which eventually becomes 18⅓. While this is likely beyond most fifth grade students, we mention this because often it is said that you <u>must</u> convert first to improper. We should instead be careful to say that converting to improper may make our calculations easier.

# Fifth Grade Math

*(Standard) long division*
- During this first math block of fifth grade, we should make the transition from *flexible long division* to *standard long division*. (See above, *Standard Long Division* under *Lessons and Topics for Fifth Grade Math*.)
- Standard long division is then practiced through the rest of fifth grade.

Review and practice! Material that has been introduced in previous blocks and in previous years needs to be reviewed, practiced, and furthered. (See above, *Practice and review of old material* under *Lessons and Topics for Fifth Grade Math*.)

Activities and Movement
- With the geometry lessons, many movement exercises can be done outside, such as:
    - The students can move their arms to represent different angles.
    - The class can form a circle with one student at the center and everyone else an equal distance from the center person.
    - In a similar way, the class can form different triangles, angles, squares, parallel lines. etc.
- *Making A Right Angle with a Rope.*
    - The following is an example of how the Egyptians were able to make a right angle by using a rope.
        - Bring the class into a field and have a rope that is 120 feet long, with a knot tied 50 feet from one end and 40 feet from the other end. (The distance between the two knots should now be 30 feet.) Form a triangle with the rope, such that the three corners of the triangle are the two knots and the place where the ends of the rope come together. All three sides of the triangle should be tight and straight. The result should be a right triangle.
        - You should show that if one knot is moved a few feet, then you no longer get a right angle.

Bookwork
- After practicing each geometric form on loose drawing paper, each form should be drawn beautifully in the main lesson book.
- Be sure to visit our website (www.JamieYorkPress.com). Under the "Resources" tab you can see full-color pages from students' main lesson books.
- We have also included a few sample black-and-while drawings for this freehand geometry block in the appendix.

No Lesson Plan Outline. Given the open-ended nature of this block, we did not feel the need to give a week-to-week lesson plan outline.

Fifth Grade Math

# Math Main Lesson Block #2: Decimal Fractions and the Metric System

New material and content

*Decimal Fractions.*
- We can begin with the question: what is the meaning of 'deca'? Some examples[35]: <u>deca</u>gon, <u>deca</u>thlon, <u>deca</u>de, <u>Deca</u>mber (the tenth month of the Roman calendar), <u>deci</u>mal.
- We should review our place value work in second and third grade. At that time, we saw that each step in place value (moving to the left) increased the value of the digit's place by a factor (multiple) of ten. Now, with decimal fractions, each step taken to the right decreases the value of the digit's place by a factor of ten.
- *A fun exercise with decimals.*
    - There is a hat for each place value (e.g., thousands, hundreds, tens, ones, tenths, hundredths, thousandths). One student is the decimal point. Starting out simply, we could just work with three digits. So, we line up three students and give each one a digit (but not a place value yet) – for example, the left-most student gets a 7, the next student gets a 1, and the last student gets a 5.
    - The 'decimal point' student then positions herself between two of the digits – assume for now between the 7 and the 1. Another student then takes the collection of hats, and gives the 'ones' hat to the student holding the 7, the 'tenths' hat to the student holding the 1, and the 'hundredths' hat to the student holding the 5. The class then reads it as "seven and fifteen hundredths".
    - We can then collect just the hats, and ask what happens if the decimal place moves one place to the left or right. The hats then get reassigned and the class reads the new number.
- *Fluency and practice.* The students should quickly become comfortable with decimal fractions.
    - We can say to the students: "Make this number 10 times bigger, or 100 times smaller, etc."
    - The students need to become comfortable when there are several digits behind the decimal point, and should be able to answer questions like: "What is bigger 1.78 or 1.76993?"
    - We should also practice rounding decimals.
    - We should show the students how most monetary systems are based on the decimal system.
    - We can bring digital time in relation to sporting events.
- *The four processes with decimal fractions.*
    - For vertical addition and subtraction of decimal fractions, place value is very important. Therefore, we need to make sure that we line up the decimal point.
    - We should also practice the four processes with decimal fractions orally (mental arithmetic) and in written horizontal form as well (e.g., $3.25x10 = __).
    - The division of decimal fractions could briefly be introduced at the end of fifth grade, or it could wait until sixth grade.
- *Simple equivalences.*
    - The students should discover that, for decimal fractions, we need to find equivalent fractions in tenths, hundredths, and thousandths. Some examples are:
        $0.3 = {}^3/_{10}$    $0.81 = {}^{81}/_{100}$    One half = ½ = 5/10 = 0.5    $0.25 = 25/100 = ¼$
- More extensive practice of converting common fractions to decimal fractions is done in the next math block, and in sixth grade.

---

[35] Dorothy Harrer, *Math Lessons For Elementary Grades.*

# Fifth Grade Math

*The Metric System.*
- This is the introduction to the metric system as a universal measurement language.
- *For those people not in the USA.* If the metric system is the norm in your country, then you can still use this opportunity to show in depth how the metric system uses decimals, and have the students work on making conversions within the metric system – e.g., "5.3cm equals how many meters?"
- The debate on whether the United States should adopt the metric system has been going on for nearly 200 years. The United States is one of the few countries in the world (Liberia and Burma being the others) that hasn't committed to adopting the metric system.
- In contrast to the U.S. measurement system, the metric system eliminates the need to deal with common fractions. The metric system is considered to be a simpler form of measurement. It is based on the decimal system (units of ten).
- With all three types of measurement (linear measurement, weight, and volume/capacity), these are the things to keep in mind:
    - We should interweave the introduction to the metric system with a review and furthering of the U.S. system. The students should become comfortable with both the U.S. and the metric units.
    - The students should measure a great many objects using the metric system, and experience measuring and estimating in both systems.
    - Estimating is key! Don't forget to estimate each time before you measure.
    - We should *not* have the students practice converting between the U.S. system and the metric system (e.g., 7ft = __ m) – this is theme for eighth grade.
- *Linear Measurement.*
    - In order to make it practical, we can begin, as we did in third grade, with measuring things in the classroom. We can measure students' heights, the perimeter of books, the lengths of desks, the blackboard, and the whole classroom. We can even measure the campus. We can use the students' rulers, meter sticks, and tape measures. We should measure both in meters and in centimeters.
    - From all of our measuring work, we can then introduce the other units of linear measurement: millimeter (mm), centimeter (cm), decimeter (dm), meter (m), decameter (dam), hectometer (hm), kilometer (km). The students need to get a sense that every step is a factor of ten.
    - Out of the practical work, the students should become comfortable with both the U.S. units (inches, feet, miles) and the metric units (cm, m, km).
    - *Simple conversions within the metric system.* The students should practice converting metric units (e.g., 7m = __ cm). Keep it simple. Much more of this will be practiced in the coming years.
- *Weight.*
    - The students should have many experiences with weighing objects, first in the U.S. system, and then in the metric system. It should be practical. Always remember to estimate before weighing!
    - Out of the practical work, the students should become comfortable with the units in both systems:
        - U.S. system: ounces, pounds, tons.
        - Metric system: milligram (mg), centigram (cg), decigram (dg), gram (g), decagram (dag), hectogram (hg), kilogram (kg).
    - *Simple conversions within the metric system.* The students should practice converting from one unit to the other (e.g., 18kg = __ g).
- *Capacity (volume).*
    - Once again, the students should have many experiences measuring the capacity of objects first in the U.S. system, and then in the metric system (e.g., fill up bottles, cans, tins, etc.). Estimate!
    - The students should become comfortable with the units in both systems:
        - U.S. system: teaspoon, tablespoon, fluid ounce, cup, pint, quart, and gallon.
        - Metric system: milliliter (mℓ), centiliter (cℓ), deciliter (dℓ), liter (ℓ), decaliter (daℓ), hectoliter (hℓ), kiloliter (kℓ).

Review and practice! Material that has been introduced in previous blocks and in previous years needs to be reviewed, practiced, and furthered. (See above, *Practice and review of old material under Lessons and Topics for Fifth Grade Math*.)

Activities and Movement
- Lots of hands-on work!
- There are many exercises that the teacher can create. (See also "*A fun exercise with decimals*" above.)
- Move the various distances!
  - Have everyone in a line and then move forward 1 centimeter.
  - Have a jumping contest where the winner is the person that jumps closest to exactly 1 meter!
  - Have the whole class run a distance of one kilometer.

Bookwork
- The students should make charts, tables, and graphs of measurements (estimations and actual). As always, the illustrations should be done beautifully.
- Be sure to visit our website (www.JamieYorkPress.com). Under the "Resources" tab you can see full-color pages from students' main lesson books.

## Lesson Plan Outline for Block #2 – Decimal Fractions and the Metric System

To be done at least occasionally all year:
- Simple word problems.
- Puzzles and games (see our puzzle/game book).
- Mental math or *Arithmetic Facts Speed Sheets* (one or the other each day).
- Vertical arithmetic (+/-/x/÷) – a bit each day!
- Practice all skills with (common) fractions – a bit each day!
- Estimation.

Week #1
- Introduction to decimal fractions
- Review place value.
- Do "A fun exercise with decimals" (see above).

Week #2
- Making a decimal fraction 10 times bigger, or 100 times smaller, etc.
- What is bigger 1.78 or 1.76993?
- Monetary systems and the decimal system.
- Digital time in relation to sporting events.
- The introduction to the metric system as a universal measurement language.
- Intro to metric linear measurement (including mm, cm, m, km). Also, simple conversions within the metric system (e.g., 7m = ___ cm).
- Review and furthering of the U.S. measurement system.

Week #3
- Practice rounding decimal fractions.
- Introduce addition and subtraction with decimal fractions.
- Simple equivalences, such as: $0.3 = {}^3/_{10}$ ; $0.25 = {}^{25}/_{100} = ¼$
- Intro to metric weight (including mg, g, kg). Also, simple conversions within the metric system (e.g., 18kg = ___ g).

Week #4
- Wrap up the block.
- Practice addition and subtraction with decimal fractions.
- Introduce multiplication with decimal fractions.
- Intro to metric volume (including mℓ, ℓ). Also, simple conversions within the metric system (e.g., 400mℓ = ___ ℓ).

Fifth Grade Math

# Math Main Lesson Block #3: The Wonder of Number

Notes:
- This is a truly wonderful main lesson to bring to the children as long as the teacher has the time and desire to learn and penetrate the material sufficiently. None of it is compulsory. Whatever isn't covered in fifth grade, could be covered at any point in grades six through eight, or not covered at all.
- A good amount of what is listed here is for the teacher's curiosity and interest. It should only be brought into the classroom if the teacher has adequately penetrated the material, and can find an appropriate and creative way to bring it to the children.
- *Wonder arises from discovery!* It is important that the students discover for themselves the properties of numbers described below, or at least, that the students come to some understanding of where these properties come from.
- This block can be either three or four weeks long. The lesson plan outline (see below) is based upon three weeks. If you have four weeks, then more fifth grade skills topics can be included.
- *Review and practice!* Material that has been introduced in previous blocks (including decimal fractions) and in previous years still needs to be reviewed, practiced, and furthered.

"Theorem Hunting"
- The focus of much of this block is to find hidden properties or theorems. Of course, the word "theorem" plays an important role in higher level mathematics. We can introduce the idea of a theorem to fifth graders in a way that isn't formal or intimidating. We can simply say:

    *A theorem is a special law in mathematics that is true in general, is interesting, and perhaps even surprising.*
- "5+9=14" is true, but it's not a theorem (it's only a fact) because it's not a statement about numbers in general – and it's not really so interesting, either.
- "Any four-digit number is greater than any two-digit number" is a general statement, but it's not interesting enough to call it a theorem.
- There are many theorems (general, interesting, and even surprising) found below.
- We will be hunting for theorems ourselves during this block.

Divisibility Rules
- *The concept of divisibility*. We need to make sure that the students are clear about what is meant when we say something like: "252 is divisible by 9." First of all, even though we can technically divide any number by any other number (and likely get a remainder), when we say one number is *divisible* by another number, it implies that it is *evenly* divisible, which means that there is no remainder after the process of division takes place. Therefore, the statement "252 is divisible by 9" guarantees that there will be no remainder if I divide 252 by 9.
- *Four questions.* The students should understand that each of these questions are essentially asking the same thing:

    *Is 74 divisible by 3?*
    *Do I get a remainder of zero with 74÷3?*
    *Is 74 a multiple of 3?*
    *Is 74 in the 3's table?*        (In every case the answer is "no".)

    *Is 732 divisible by 4?*
    *Do I get a remainder of zero with 732÷4?*
    *Is 732 a multiple of 4?*
    *Is 732 in the 4's table?*        (In every case the answer is "yes".)
- Here are the divisibility rules that can be done in fifth grade – each one expressed as a theorem:
    - A number is <u>divisible by 10</u> (i.e., can be divided by 10 with no remainder) if the number ends in 0.
    - A number is <u>divisible by 5</u> only if the number ends in 5 or 0.
    - A number is <u>divisible by 2</u> only if it is even.

# Fifth Grade Math

- A number is <u>divisible by 4</u> only if the last two digits are divisible by 4.
  - We can guide them to discover this by writing the extended 4's table on the board up to 400, and then allow them to discover that once we get past 100, the last two digits keep repeating: 1<u>00</u>, 1<u>04</u>, 1<u>08</u>, 1<u>12</u>, etc., and then 2<u>00</u>, 2<u>04</u>, 2<u>08</u>, 2<u>12</u>, etc., and then 3<u>00</u>, 3<u>04</u>, 3<u>08</u>, 3<u>12</u>, etc.
  - This is particularly fun for the children because, after they have learned it, someone can give them any large number and they only have to listen for the last two digits in order to determine if the number is divisible by 4. The students can then try to figure out how to do this quickly in the case that the last two digits are greater than 50. One method is to subtract 20, 40, 60 or 80 (which are all divisible by 4) in order to make it easier.
  - <u>Example</u>: Is 6,380,716 divisible by 4? It ends in 16, which is divisible by 4. Therefore, 6,380,716 is also divisible by 4.
  - <u>Example</u>: Is 98,273,474 divisible by 4? We see that it ends in 74. We subtract 60, leaving us with 14, which isn't divisible by 4. Therefore, 98,273,474 is *not* divisible by 4.
- A number is <u>divisible by 9</u> only if the sum of the digits is divisible by 9.
  - Here is one way to guide the students to discover this for themselves. Have the students in groups write the extended 9's table all the way up to 300. Ask them: "What do you notice about all of the numbers in the 9's table?" Someone may say that the digits always add to 9. But then we see some exceptions, such as 99, 189, 288. Then ask them to choose any number that has three, four, or five digits, and multiply that number by 9. Write all the results on the board. Each number on the board must be in the 9's table. What do you notice? (Perhaps we leave the question hanging overnight.) Eventually, they see that any number in the 9's table has digits that add to a sum that is also in the 9's table! This can then be reworded (as given above) as: "a number is divisible by 9 only if the sum of the digits is divisible by 9".
  - <u>Example</u>: Is 6,507 divisible by 9? The sum of the digits is 18, which is divisible by 9. Therefore, 6,507 is divisible by 9.
- A number is <u>divisible by 3</u> only if the sum of the digits is divisible by 3.
  - <u>Example</u>: Is 71,284 divisible by 3? The sum of the digits is 22, which isn't divisible by 3. Therefore, 71,284 is not divisible by 3.

<u>Discovering the Prime Numbers</u>. (See also *Prime Numbers up to 2000* in the appendix)
- Start with reviewing factors. (See factors from fourth grade.)
- The idea of prime numbers should have been introduced in third or fourth grade, but now in fifth grade, we create a list of the prime numbers. After reviewing the idea of a prime number, we can break the class into small groups and simply say: "Write a list of prime numbers, starting at 2, and going as far as you can." This activity can be equally engaging for students at all ability levels. Some students may only go up to 50 (which is fine), while the most advanced students may even try to go up to 500.
- Invite each student to write one of their prime numbers on the board. In the end, the list should (at least) include all the prime numbers up to 100. Besides missing a couple of primes, the list of primes that the students have put on the board might mistakenly include a few composite numbers (numbers that aren't prime). Often this can be something like 87 (which is divisible by 3). The most difficult number to recognize as composite is 91 (which is 7 times 13). It's good to leave some mistakes on the board overnight, and then correct the mistakes the next day. Leave the (correct) list of prime numbers on the board through the rest of the main lesson, since it will be used in the coming days.

<u>Powers of Two</u>. (See also *Powers of Two Table* in the appendix.)
- We can ask the children: What do we get if we multiply two 2's together? How about multiplying three 2's together? And ten 2's? And twenty 2's? They may be surprised with how quickly the results get large.
- It may be best to not introduce exponent notation until sixth grade.
- Students who need an extra challenge can make a list of the powers of three, four, or five. They may be

90

## Fifth Grade Math

amazed to discover that the powers of four are found by taking every other number from the list of the powers of two.

Square and Triangular Numbers. (See also *Square and Triangular Numbers* in the appendix.)
- *The square numbers* are 1, 4, 9, 16, etc.
  - They are found by multiplying each counting number times itself: 1x1, 2x2, 3x3, etc.
  - However, it is best to have the students discover these numbers as the ancient Greeks did – from actual squares. They simply place stones (or dots) into square shapes: three rows of three make 9, four rows of four make 16, etc., as shown here.

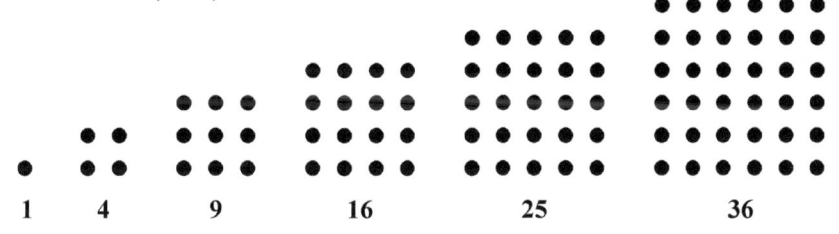

     1    4    9    16    25    36

- *The triangular numbers* are 1, 3, 6, 10, 15, 21, 28, 36, etc.
  - The reason that they are called triangular can be understood by looking at the way the dots are placed in a triangular form, as shown here.

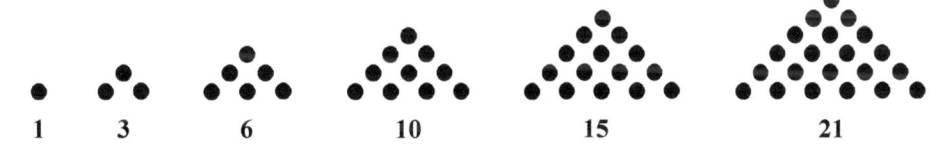

     1    3    6    10    15    21

- *Discovering theorems!* After the students have discovered the square and triangular numbers, we invite them to discover some theorems. We simply ask them: "What patterns and properties do you see with these numbers?" Here are some possible theorems involving square and triangular numbers:
  - "The differences of the square numbers are the odd numbers."

    Square Numbers:  1    4    9    16    25    36...
                            3    5    7    9    11  Odd Numbers

  - "The differences of the triangular numbers are the counting numbers."
  - "Every square number is the sum of the first several odd numbers."
    For example, 16=1+3+5+7, and 36=1+3+5+7+9+11.
  - "Every triangular number is the sum of the first several counting numbers."
    For example, 10=1+2+3+4, and 21=1+2+3+4+5+6.
  - *A neat theorem!* "The sum of two consecutive triangular numbers is always a square number." This can be quite a thrill for students to discover on their own! A nice "proof" of this is shown with the drawing here. 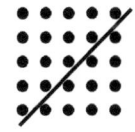
  - *A very surprising theorem![36]* "If we write the first n square numbers in descending order, with – and + signs alternating between them (starting with –), the result is always the $n^{th}$ triangular number."
    For example, 16–9+4–1 = 10 and 49–36+25–16+9–4+1 = 28

- Have the students try to find the first three numbers that are *both* square and triangular. There are only seven numbers below 2 billion that are *both* square and triangular. They are
      1   36   1225   41,616   1,413,721   48,024,900   1,631,432,881

---

[36] This theorem was discovered by two fifth grade students (Forest Lytle and Raven Armour) at Shining Mountain Waldorf School in 2017.

# Fifth Grade Math

<u>Perfect, Abundant, and Deficient Numbers</u>.  (See *Perfect, Abundant, and Deficient Numbers* in appendix.)
- The ancient Greeks (especially the Pythagoreans) believed that certain numbers had special meaning and significance. They studied perfect, abundant, and deficient numbers in detail.
- *What are perfect, abundant, and deficient numbers?*  Starting at 2, each counting number is categorized as abundant, deficient or perfect based upon the sum of its factors. In order to determine whether a given counting number is perfect, abundant or deficient, we first list all the number's factors, <u>except for the number itself</u>. Then we sum up the numbers in that list. If this sum is equal to the number itself, then we say the number is *perfect*. If this sum is less than the number itself, then we say the number is *deficient*. If the sum is greater than the number itself, then the number is *abundant*.
- *Discovering Perfect Numbers.*[37]  Show the students that 10 is a *deficient number* because the sum of its factors (1, 2, 5) is less than 10. Then show the students that 20 is an *abundant number* because the sum of its factors (1, 2, 4, 5, 10) is greater than 20. Then tell the students that there are only two *perfect numbers* under 60, and let them try to find these two special numbers. (They are 6 and 28.)
- *More perfect numbers?*  Perfect numbers are extremely rare; there are only four perfect numbers below 10,000,000! The students now know that the first two perfect numbers are 6 and 28. You could then give the students the next two perfect numbers (496 and 8128) and have them find all the factors of each one (perhaps with the clue that half the factors are powers of two). They then need to prove that it is perfect by showing that the sum of the factors (except for the number itself) is equal to the number itself. You can then tell the students that they will have to wait until seventh grade to learn how to calculate more discover numbers. (A bit of drama never hurt!)
- *The abundance quotient* is the quotient that results when the sum of the factors (without the number itself) is divided by the number itself.

  <u>Example</u>:  Determine whether the number 10 is perfect, abundant or deficient, and calculate its abundance quotient.
  <u>Solution</u>:  The list of factors for 10 is 1, 2, 5. The sum of these factors is 8, which is less than the number itself (10), so the number is *deficient*. The abundance quotient is 8÷10, which is <u>0.8</u>.

  <u>Example</u>:  Determine whether the number 20 is perfect, abundant or deficient, and calculate its abundance quotient.
  <u>Solution</u>:  The list of factors for 20 is 1, 2, 4, 5, 10. The sum of these factors is 22, which is greater than the number itself (20), so the number is *abundant*. The abundance quotient is 22÷20, which is <u>1.1</u>.

  <u>Example</u>:  Determine whether the number 28 is perfect, abundant or deficient, and calculate its abundance quotient.
  <u>Solution</u>:  The list of factors for 28 is 1, 2, 4, 7, 14. The sum of these factors is 28, which is equal to the number itself (28), so the number is *perfect*. The abundance quotient (for all perfect numbers) is exactly <u>1</u>.

- *More about perfect, abundant, and deficient numbers.*
  - There are 21 even abundant numbers under 100. The first odd abundant number is 945.
  - The first 7 perfect numbers are:
    6;  28;  496;  8128;  33,550,336;  8,589,869,056;  137,438,691,328.
  - Early on, the Greeks knew that the first three perfect numbers were 6, 28, and 496. They had troubles finding perfect numbers beyond that because the numbers were getting too large to list and add all the factors. They wanted to find an easier method for determining perfect numbers.
  - Euclid discovered a formula for calculating even perfect numbers around 300B.C. This formula (See *A Source Book for Teaching Middle School Math, A Middle School Math Curriculum*, under seventh grade algebra) made it possible to discover the next few perfect numbers (perhaps up to the seventh one). In the 1600's, Jean Prestet found the eighth perfect number: 2,305,843,008,139,952,128.
  - There are no known odd perfect numbers, and it is one of the great mysteries of mathematics whether or not an odd perfect number could possibly exist.

---

[37] Note that 1 is not considered to be a perfect number for the same reason that it isn't considered to be a prime number: it is the basis of all numbers. Also, our list of factors isn't supposed to include the number itself, so the number 1 doesn't have any numbers in its list of factors.

# Fifth Grade Math

## Sum and Difference Theorems

- Each of the below theorems expresses some surprising law about the relationship of numbers. If properly brought to the children, this can engender a real sense of wonder. It should only be brought to the children if the teacher has developed a good connection to this material.
- The challenge is to not make it too abstract. Try to make it playful.
- The ancient Greeks were very interested in how numbers could be expressed as the sum or difference of other special numbers (e.g., prime numbers or square numbers).
- For the below theorems, it is best to have a list of the prime numbers and a list of the square numbers (which the students discovered earlier in the block) on the board for reference.
- Around 1640, the famous French mathematician, Pierre de Fermat, came up with the below theorems (except for the first one):
- *Goldbach's Conjecture*: *"Every even number can be expressed as the sum of two prime numbers."*
    - Simply ask the children how many ways they can find to express 90 as the sum of two prime numbers. Write all the different possibilities that they have found on the board. They will be amazed to see that there are nine ways to express 90 as the sum of two primes (7+83; 11+79; 17+73; 19+71; 23+67; 29+61; 31+59; 37+53; 43+47).
    - In contrast, there are only two ways to express 68 as the sum of two primes (7+61 and 31+37).
    - Once we get past the first few even numbers, most all of them can be expressed as the sum of two prime numbers in multiple ways, yet the number of possible ways varies greatly.
    - While most mathematicians believe that *Goldbach's Conjecture* is true, nobody has been able to prove it (but many have tried!), which is why it is a *conjecture*, and not a *theorem*.
    - See the appendix, *Even Numbers as the Sum of Two Primes,* for a list of all the ways that the even numbers from 4 to 150 can be expressed as the sum of two primes.

- *The difference of two square numbers.*

  *"Every prime number, except for 2, can be expressed as the difference of two square numbers in one and only one way."*
    - The students should first make a list of consecutive square numbers:
        1  4  9  16  25  36…   We see that the differences/distances grow.
           3  5  7  9  11…   ← These are the differences between neighbors.

    Given that the above differences between neighbors form the list of odd numbers, we can easily see how *any odd number can be expressed as the difference of two squares*. For example, we can express 7 as $16-9$. Similarly, we can express 15 as $64-49$.

    The real surprise with the above theorem is that it says, if the odd number is a prime number, then it can be expressed as the difference of two squares in *only one way*. For example, 15 can be expressed as the difference of two squares *in two ways*: either as $64-49$ or as $16-1$. This does not contradict our theorem because 15 is not a prime number. 7, on the other hand, is prime, so we know that it can only be expressed as the difference of two squares in one way: $16-9$.

    As another example, consider the numbers 1047 and 1049. How many ways can each of them be expressed as the difference of two square numbers? We look at the table of prime number (see the appendix) and see that 1049 is prime, and 1047 (which is divisible by 3) is composite. The theorem therefore guarantees that 1049 (which is prime) can be expressed as the difference of two squares in only one way (which turns out to be $525^2 - 524^2 = 275625 - 274576 = 1049$). On the other hand, 1047 can be expressed as the difference of two squares as $524^2 - 523^2$, but we aren't immediately sure if 1047 (which isn't prime) can be expressed as the difference of two squares in another possible way, or not.
    - See the appendix, *Odd Numbers as the Difference of Two Squares*, for a list of all the ways that the odd numbers from 3 to 299 can be expressed as the *difference* of two squares.

# Fifth Grade Math

- *The sum of two square numbers.* It is a huge challenge to guide fifth grade students to discover the following theorems – but it is possible!

  *"If a number is prime and has a remainder of 1 after dividing it by 4, then it can be expressed as the sum of two square numbers in one and only one way."*
  - Example: The number 73 is both prime and has a remainder of 1 when divided by 4. This theorem tells us that there must be exactly one way to express 73 as the sum of two squares. Looking in the appendix (at *Numbers as the Sum of Two Squares*), we see that 73 can be expressed only as 64 + 9.
  - See the appendix, *Numbers as the Sum of Two Squares*, for a list of all the ways that the numbers from 2 to 442 can be expressed as the *sum* of two squares.

  *"If a number is prime and has a remainder of 3 after dividing it by 4, then it is not possible to express it as the sum of two square numbers."*
  - Example: The number 43 is both prime and has a remainder of 3 when divided by 4. Therefore, according to the theorem, it must be impossible to express 43 as the sum of two squares. Looking in the appendix (at *Numbers as the Sum of Two Squares*) confirms that 43 can't be expressed as the sum of two squares.

  *"If a number is not prime, then there are a variety of possibilities – it may be that the number can be expressed as the sum of two square numbers in one way, in multiple ways, or not at all."*
  - Examples: (From the appendix, *Numbers as the Sum of Two Squares*)
    - 45 can be expressed as the sum of two squares in exactly one way: 36 + 9.
    - 48 can't be expressed as the sum of two squares in any way.
    - 50 is the first number that can be expressed as the sum of two squares in 2 ways:
      25 + 25  or  49 + 1
    - 325 is the first number that can be expressed as the sum of two squares in 3 ways:
      1 + 324;  36 + 289;  100 + 225
    - 1105 is the first number that can be expressed as the sum of two squares in 4 ways:
      16 + 1089;  81 + 1024;  144 + 961;  529 + 576

Fifth Grade Math

# Lesson Plan Outline for Block #3 – The Wonder of Number

Note: Be sure to integrate review of topics covered in earlier blocks, especially decimal fractions and common fractions.

Day #1
- Tell the story of Pythagoras – how he made some of the first important mathematical discoveries.
- Talk about the idea of a "theorem", and about how we will be discovering theorems.
- Review factors, including finding all the factors of a given number.

Day #2
- Introduce the idea of divisibility, including the "four questions".
- State the divisibility rules for 2, 5, and 10 as theorems (e.g., "A number is divisible by 5 only if the number ends in a 5 or a 0.").
- Give a few large numbers and ask if they are divisible by 2, 5, and 10.

Day #3
- Build up to the discovery of the divisibility rules for 4 and 9.
- Review the idea of a prime number.

Day #4
- Ask: "How do we know if a number is divisible by 3?", then state that rule as a theorem.
- Summarize the divisibility rules (theorems) that we know so far.
- In small groups, have the students find as many prime numbers as they can – up to 100, or more.

Day #5
- Consolidate all that has been done this week.

Day #6
- Finish creating a list of prime numbers up to 100.
- Introduce the idea of summing the factors of a number to see if it is deficient or abundant.

Day #7
- Do more with deficient and abundant numbers.
- Ask: what happens with a prime number? Answer: It's deficient because the factor sum (not including the number itself) is always 1.
- Challenge: Can you find a number that has a factor sum that is twice the number itself? Ans: 120

Day #8
- Introduce the idea of a perfect number, and then break into small groups and try to find the first two perfect numbers. Give hint that they are less than 60.

Day #9
- In small groups, show that 496 is a perfect number.
- Challenge: show that 8128 is perfect.

Day #10
- Give the first eight perfect numbers, and tell some of the interesting facts of perfect numbers.
- Introduce triangular and square numbers. In small groups, try to find as many as possible.

Day #11
- With the first 15 square and triangular numbers on the board, ask: "What patterns or theorems do you see?"
- *Goldbach's Conjecture.* While looking at the list of primes on the board, give the students a few even numbers and have them find multiple ways to write each one as the sum of two primes.

Day #12
- Do a bit more with *Goldbach's Conjecture.*
- *Sum of Two Squares.* With the first 25 square numbers written on the board, give the students several numbers (13, 29, 17, 18, 14, 50, 31, 37, 130, 157, 119, 325, etc.) to write as the sum of two squares. Some can be done in only one way, some in two ways, and some not at all. Mention that there is a very surprising theorem behind this.

Day #13
- Continue work with the *Sum of Two Squares.*
- Create three lists on the board: those numbers that can be expressed as the sum of two squares in multiple ways, those that can be expressed in only one way, and those that are impossible.
- State that the clue to discovering the theorem is the prime numbers. Circle the prime numbers in each list. What do you notice?

Day #14
- Bring the *Sum of Two Squares* to a close. State the theorems associated with this.
- Begin to wrap up the block.

Day #15
- Bring the block to a close, and summarize/review all that we have learned and experienced.

# Appendix

Appendix

# A Math Curriculum Summary for Grades One through Five

## — First Grade —

*The world of numbers.*
- Roman numerals. Begin with the Roman numerals, and then shortly thereafter, introduce the (standard) Arabic numerals.
- Quality of numbers. What is the quality of the numbers in the surrounding world?
- Counting. Counting forward and backward comfortably up to 100. Awaken a sense for number.
- Number dictations. There should be number dictations starting in the second math block.

*Developing a sense of numbers.*
- Movement. The challenge is to synchronize the movement with the speaking voice.
- Estimating. "How many steps am I from the board?"
- Rhythmical and skip counting with the 2's, 3's, 4's, 5's, and 10's. This serves as preparation for learning the times tables in second grade.

*Beginning calculations.*
- All four processes are introduced. The children become comfortable with adding and subtracting numbers up to 24, and fluently up to 10.
- Regrouping numbers. For example, how can we regroup the number ten?
- Learning the "easy" addition facts. All addition facts up to 10, as well as all of the doubles (i.e., 6+6, 7+7, 8+8, 9+9) should be learned by heart by the end of first grade.

## — Second Grade —

*Place value.* Place value should be introduced and practiced. This is an important step.

*The world of numbers.* The students should become comfortable with numbers up into the 1000's.

*Estimating.* We build up from the estimating done in first grade and progress to more challenging estimations.

*Addition and subtraction facts.* By the end of the year, the class should learn *by heart* their addition facts (up to 9+9=18) and the corresponding subtraction facts.

*The times/division tables.* By the end of the year, the class should be comfortable with all of the times/division tables from 2-12, in a row (e.g., for the 7's table: 7, 14, 21, 28, etc.).

*The four processes.*
- *Horizontal only!* We are not yet working with vertical procedures (e.g., carrying, borrowing, etc.)
- *Addition.* By the end of the year, the class should become comfortable with adding any two-digit number with a one-digit number (e.g., 57+6).
- *Subtraction.* By the end of the year, the class should become comfortable with subtracting any one-digit number from a two-digit number (e.g., 52–6), and also with subtracting two 2-digit numbers such that the answer is a one-digit number (e.g., 72–69).
- The students must gain an understanding of the concept of multiplication and division, such as:
  $3 \times 2 = 6$ means: "three groups of two make six."
  $12 \div 3 = 4$ means: "three fits into twelve four times."
- By the end of the year, the children should be able to do all four processes (even alternating on the same page) and know the difference between the processes without help. But keep the problems simple!

*Time orientation.* The days of the week and the months of the year.

*The wonder of number.* Geometrical patterns that arise from the times/division tables and the circle.

Appendix

## — Third Grade —

*The world of numbers.* Fluent with numbers in the 1000's and comfortable with the numbers in the millions.

*Learning all of the arithmetic facts!!* Now is the time to become fluent with the multiplication/division facts (that come from the 2-12 tables) out of order, as well as all of the addition facts and subtraction facts. Systematic daily work, both orally and written (with our *arithmetic facts practice sheets*) is needed.

*The four processes – working vertically.*
- An introduction to vertical addition ("carrying") and vertical subtraction ("borrowing"). Build up to adding two 4-digit numbers (e.g., 8364+8375) and subtracting two 3-digit numbers (e.g., 643–387).
- An introduction to vertical multiplication. Only do single-digit multipliers (e.g., 2347x5).
- An introduction to long division should wait until fourth grade.

*The four processes – working horizontally.* Even though the children are being introduced to working with the four processes in vertical form, the bulk of their work with the four processes is still in horizontal form.

*Measurement.* Introduction to time, distance, weight, volume, and money (currency).

## — Fourth Grade —

*The world of numbers.* GCF (greatest common factor) and LCM (least common multiple).

*The arithmetic facts.* Continued work both orally and written (e.g., with our *arithmetic facts review sheets*).

*The four processes.* Regular practice is needed.
- *Horizontal addition and subtraction.* Problems like: 125+126, 895+112, 974–875.
- *Vertical addition and subtraction.* Work with larger numbers than what was done in third grade.
- *Vertical multiplication.* We can build up to three-digit multipliers (e.g., 4372x836).
- *An introduction to flexible long division.* Build up to four-step problems. The divisors should be kept between 2 and 12 (e.g., 15288÷6=2548). (*Standard long division* waits until fifth grade.)

*Fractions.* An introduction to fractions, including: types of fractions, equivalent fractions, and simple arithmetic with fractions. The goal is to get to the end of fourth grade with the class having a good understanding of the basics of fractions, and having the children saying, "I like fractions!"

*Measurement.* Review third grade measurement, and begin doing simple conversion problems.

## — Fifth Grade —

*The Wonder of Number.* Some of the possible topics are: *Square and Triangular Numbers, Powers of Two, Divisibility Rules, Perfect Numbers,* and *Sum and Difference Theorems.* Leave them in wonder!

*Geometry.*
- *Freehand geometry.* The students create many beautiful drawings of circles, squares, triangles, angles, divisions of the circle, etc., all freehand, without the aid of a ruler or compass.
- We can bring an imaginative picture of the *Pythagorean Theorem*.
- We can also introduce *perimeter and area*.

*(Common) Fractions.* The students should be fluent with adding/subtracting fractions with like denominators and comfortable with adding/subtracting fractions with unlike denominators. Multiplying and dividing of common fractions should be thoroughly practiced.

*Decimal fractions.* The number world of decimal fractions is introduced.

*Measurement.* The metric system (distance, weight, and capacity) is introduced. Review of the U.S. system is integrated into our work with the metric system.

*Review and Practice.* It is especially important to review and further all work with vertical arithmetic (addition, subtraction, long multiplication, and long division), and all work with fractions.

*Arithmetic facts.* Review and practice of the arithmetic facts is still important. This can be done a couple of times per week both orally and written (e.g., with our *arithmetic facts speed sheets*).

Appendix

# A List of Games for Lower Grades Math

Notes:
- All of the following games[38] are related to math.
- It is fun for the students to experience the world of numbers in a playful way.
- All these games can be used from grade two and up.

### Our puzzle and game book
- There may be times when the teacher knows it is time to do something different. Our book *Fun with Puzzles, Games, and More!* is intended as a resource for math teachers in grades four through twelve, in part to supplement the normal classroom material. It provides ideas for that "something different". It can be ordered through our website www.JamieYorkPress.com.

### Dominoes and card games (there is a great variety)
- Krypto Card Games
- Numbers Factory

### Bingo games (there is a great variety)
- Conceptual Bingo with Fractions
- Conceptual Bingo with Decimals
- Conceptual Bingo with Fractions to Decimals

### Traditional games (that can help with simple counting, and arithmetic)
- Monopoly
- Yahtzee
- Shut the Box
- Can't Stop

### Games using the four processes (+,−,x,÷)
- Equate (All Star Math games)
- Roll 'n Multiply (All Star Math games)
- Smath: The Game That Makes Math Fun.
- Hands on Learning, math card games
- Math Mania Highlights

### Other ideas
- A Box with Six Math Board Games for all Ages
- Mind Bending Puzzles Books, by Lagoon Books
- Books with Mazes
- Books with Patterns

---

[38] Most of these games were found at the store "Math and Stuff" in Seattle. It is a wellspring of math resources. Visit their website at www.math-n-stuff.com.

Appendix
# A Step-by-Step Progression for the Arithmetic Facts

**Step #1:** *Basic rhythmical counting and skip counting* (first grade). In first grade, rhythmical counting and skip counting exercises are done with the 2's, 3's, 4's, 5's, and 10's tables.

**Step #2:** *Addition and subtraction facts* (first and second grade).
Starting in first grade with some of the more basic facts (e.g., 5=3+2, and 12=6+6), the children begin to learn the addition facts, and corresponding subtraction facts, by heart. By the end of second grade, these more basic facts should be solid with most of the students, and by the end of third grade, the whole class should have mastered all of the addition and subtraction facts.

**Step #3:** *Basic times/division tables* (second grade) – the 2's, 3's, 4's, 5's, 10's, and 11's tables.
Building on the rhythmical counting and skip counting exercises in first grade, in second grade, we introduce the times/division tables as a concept, in an artistic way. These are introduced through movement in different ways and also written down. We should find ways to integrate the times/division tables together. For example, we can say:       1x5=5,    5=1x5,    5÷1=5;
                                                                      2x5=10, 10=2x5, 10÷2=5;
                                                                      3x5=15, 15=3x5, 15÷3=5, etc.

**Step #4:** *Advanced times/division tables* (second grade) – the 6's, 7's, 8's, 9's, and 12's tables.
In first grade, counting exercises were not done for the 6's through the 9's. Now in second grade, they can work on the 6's table through the 9's table, and the 12's table as well. As opposed to the rhythmical counting that was done in first grade, the class can jump right to skip counting with these more advanced tables (e.g., with the 6's, we say: "6, 12, 18, 24, 30, etc.").

**Step #5:** *Written practice* (second grade).
After a times table has been introduced and we have practiced it through movement, we then need to have the children practice writing the times/division tables in a row (e.g., for the 7's table: 7, 14, 21, etc.). Each student should be able to write down each times table in a horizontal row without any help.

**Step #6:** *Arithmetic facts* (third grade).
The goal is that, by the end of third grade, the whole class has learned the arithmetic facts by heart. Our *arithmetic facts practice sheets* are designed to be done almost daily for 20 weeks. (For more details, see *All about the Third Grade Arithmetic Facts Practice Sheets* under *More Ideas for Teaching Third Grade Math*.) In third grade, we should also work with remainders (e.g., 20÷6 = 3 r 2).

**Step #7:** *Review and practice* (fourth and fifth grade).
Arithmetic facts practice, review and speed sheets should be done in fourth and fifth grade in order to reinforce what has been learned, and to keep it fresh. Two or three sheets per week – each sheet taking only 3-5 minutes – should be enough. Daily mental arithmetic also supports this effort.

## Additional notes on learning the arithmetic facts

- There may be a significant overlap between some of the above steps – i.e., one step does not necessarily need to be finished with all tables before beginning the next step with another table.
- Rhythmical work should continue in third grade, but we must ensure that it includes sufficient work with the "head memory". To do this effectively, we shouldn't only have the students recite the tables together as a class with choral speaking, but we should also call on individual students to give answers to facts out of sequence.
- *The multiplication facts* aren't so numerous. There is a total of 66 multiplication facts, but only 17 of them are "hard" (these are from the 6's, 7's, 8's, 9's and 12's, tables, plus 11x11 and 11x12). Even if you add in the 4's and 3's tables, that's only 30 "hard" facts to learn.
- *Division facts.* Starting in second grade, the division facts should always be done with the multiplication facts. For instance, if you are working on 8x7=56, then at the same time you can do 56÷8, and 56÷7, as well as problems like 8 x __ = 56.
- *Subtraction facts.*
  - The subtraction facts are often not given enough emphasis. In middle school, it is often the subtraction facts (e.g., 12–7) that slow students down the most when doing calculations.
  - There are 81 subtraction facts to memorize – 45 subtraction facts that are from 10 and under (e.g., 8–3) and 36 "borrowing" facts (e.g., 13–8). Practicing the subtraction facts in first through third grade is very important.

Appendix
# Sample Drawings for Fifth Grade Freehand Geometry

Note: The students' drawings should be done with beautiful colors on large paper. We hope that this small sample of drawings will inspire you to develop many more. See our website for more.

*from periphery to center*

*from center to periphery*

*from center to periphery*

*triangle*

*rectangle*

*square*

Appendix

# The First 100 Square Numbers

| | | | |
|---|---|---|---|
| #1 is 1 | #26 is 676 | #51 is 2601 | #76 is 5776 |
| #2 is 4 | #27 is 729 | #52 is 2704 | #77 is 5929 |
| #3 is 9 | #28 is 784 | #53 is 2809 | #78 is 6084 |
| #4 is 16 | #29 is 841 | #54 is 2916 | #79 is 6241 |
| #5 is 25 | #30 is 900 | #55 is 3025 | #80 is 6400 |
| #6 is 36 | #31 is 961 | #56 is 3136 | #81 is 6561 |
| #7 is 49 | #32 is 1024 | #57 is 3249 | #82 is 6724 |
| #8 is 64 | #33 is 1089 | #58 is 3364 | #83 is 6889 |
| #9 is 81 | #34 is 1156 | #59 is 3481 | #84 is 7056 |
| #10 is 100 | #35 is 1225 | #60 is 3600 | #85 is 7225 |
| #11 is 121 | #36 is 1296 | #61 is 3721 | #86 is 7396 |
| #12 is 144 | #37 is 1369 | #62 is 3844 | #87 is 7569 |
| #13 is 169 | #38 is 1444 | #63 is 3969 | #88 is 7744 |
| #14 is 196 | #39 is 1521 | #64 is 4096 | #89 is 7921 |
| #15 is 225 | #40 is 1600 | #65 is 4225 | #90 is 8100 |
| #16 is 256 | #41 is 1681 | #66 is 4356 | #91 is 8281 |
| #17 is 289 | #42 is 1764 | #67 is 4489 | #92 is 8464 |
| #18 is 324 | #43 is 1849 | #68 is 4624 | #93 is 8649 |
| #19 is 361 | #44 is 1936 | #69 is 4761 | #94 is 8836 |
| #20 is 400 | #45 is 2025 | #70 is 4900 | #95 is 9025 |
| #21 is 441 | #46 is 2116 | #71 is 5041 | #96 is 9216 |
| #22 is 484 | #47 is 2209 | #72 is 5184 | #97 is 9409 |
| #23 is 529 | #48 is 2304 | #73 is 5329 | #98 is 9604 |
| #24 is 576 | #49 is 2401 | #74 is 5476 | #99 is 9801 |
| #25 is 625 | #50 is 2500 | #75 is 5625 | #100 is 10000 |

# The First 75 Triangular Numbers

| | | | |
|---|---|---|---|
| #1 is 1 | #20 is 210 | #39 is 780 | #58 is 1711 |
| #2 is 3 | #21 is 231 | #40 is 820 | #59 is 1770 |
| #3 is 6 | #22 is 253 | #41 is 861 | #60 is 1830 |
| #4 is 10 | #23 is 276 | #42 is 903 | #61 is 1891 |
| #5 is 15 | #24 is 300 | #43 is 946 | #62 is 1953 |
| #6 is 21 | #25 is 325 | #44 is 990 | #63 is 2016 |
| #7 is 28 | #26 is 351 | #45 is 1035 | #64 is 2080 |
| #8 is 36 | #27 is 378 | #46 is 1081 | #65 is 2145 |
| #9 is 45 | #28 is 406 | #47 is 1128 | #66 is 2211 |
| #10 is 55 | #29 is 435 | #48 is 1176 | #67 is 2278 |
| #11 is 66 | #30 is 465 | #49 is 1225 | #68 is 2346 |
| #12 is 78 | #31 is 496 | #50 is 1275 | #69 is 2415 |
| #13 is 91 | #32 is 528 | #51 is 1326 | #70 is 2485 |
| #14 is 105 | #33 is 561 | #52 is 1378 | #71 is 2556 |
| #15 is 120 | #34 is 595 | #53 is 1431 | #72 is 2628 |
| #16 is 136 | #35 is 630 | #54 is 1485 | #73 is 2701 |
| #17 is 153 | #36 is 666 | #55 is 1540 | #74 is 2775 |
| #18 is 171 | #37 is 703 | #56 is 1596 | #75 is 2850 |
| #19 is 190 | #38 is 741 | #57 is 1653 | |

Appendix

# Perfect, Abundant, and Deficient Numbers

The abundance quotient is the sum of its factors (except for the number itself) divided by the number itself. For example, with 24, the sum of its factors is 36, so the abundance quotient for 24 is 36÷24 = 1.5.

All perfect numbers, by definition, have an abundance quotient exactly equal to one. The first nine perfect numbers are:

| | |
|---|---|
| 6 | 8,589,869,056 |
| 28 | 137,438,691,328 |
| 496 | 2,305,843,008,139,952,128 |
| 8128 | 2,658,455,991,569,831,744,654,692,615,953,842,176. |
| 33,550,336 | |

The tenth perfect number has 54 digits! It is still unknown if any odd perfect number exists.

Interestingly, the first 231 abundant numbers are all even numbers. The first odd-numbered abundant number is 945 (quotient = 1.032), and the second one is 1575 (quotient = 1.047). The *abundance quotients* of each of the "biggest" abundant numbers (i.e., having an abundance quotient greater than any previous number) from 6 up to 30,000 are listed below.

| | |
|---|---|
| 12 is abundant with a quotient of 1.333 | 720 is abundant with a quotient of 2.358 |
| 24 is abundant with a quotient of 1.500 | 840 is abundant with a quotient of 2.429 |
| 36 is abundant with a quotient of 1.528 | 1260 is abundant with a quotient of 2.467 |
| 48 is abundant with a quotient of 1.583 | 1680 is abundant with a quotient of 2.543 |
| 60 is abundant with a quotient of 1.800 | 2520 is abundant with a quotient of 2.714 |
| 120 is abundant with a quotient of 2.000 | 5040 is abundant with a quotient of 2.838 |
| 180 is abundant with a quotient of 2.033 | 10080 is abundant with a quotient of 2.900 |
| 240 is abundant with a quotient of 2.100 | 15120 is abundant with a quotient of 2.937 |
| 360 is abundant with a quotient of 2.250 | 25200 is abundant with a quotient of 2.966 |
| | 27720 is abundant with a quotient of 3.052 |

## Here are the abundance quotients for the abundant numbers up to 150:

```
12  has a quotient of  1.333      80  has a quotient of  1.325
18  has a quotient of  1.167      84  has a quotient of  1.667
20  has a quotient of  1.100      88  has a quotient of  1.045
24  has a quotient of  1.500      90  has a quotient of  1.600
28  has a quotient of  1.000      96  has a quotient of  1.625
30  has a quotient of  1.400     100  has a quotient of  1.170
36  has a quotient of  1.528     102  has a quotient of  1.118
40  has a quotient of  1.250     104  has a quotient of  1.019
42  has a quotient of  1.286     108  has a quotient of  1.593
48  has a quotient of  1.583     112  has a quotient of  1.214
54  has a quotient of  1.222     114  has a quotient of  1.105
56  has a quotient of  1.143     120  has a quotient of  2.000
60  has a quotient of  1.800     126  has a quotient of  1.476
66  has a quotient of  1.182     132  has a quotient of  1.545
70  has a quotient of  1.057     138  has a quotient of  1.087
72  has a quotient of  1.708     140  has a quotient of  1.400
78  has a quotient of  1.154
```

Appendix

# Powers of Two Table

2 to the 1 is 2
2 to the 2 is 4
2 to the 3 is 8
2 to the 4 is 16
2 to the 5 is 32
2 to the 6 is 64
2 to the 7 is 128
2 to the 8 is 256
2 to the 9 is 512
2 to the 10 is 1,024
2 to the 11 is 2,048
2 to the 12 is 4,096
2 to the 13 is 8,192
2 to the 14 is 16,384
2 to the 15 is 32,768
2 to the 16 is 65,536
2 to the 17 is 131,072
2 to the 18 is 262,144
2 to the 19 is 524,288
2 to the 20 is 1,048,576
2 to the 21 is 2,097,152
2 to the 22 is 4,194,304
2 to the 23 is 8,388,608
2 to the 24 is 16,777,216
2 to the 25 is 33,554,432
2 to the 26 is 67,108,864
2 to the 27 is 134,217,728
2 to the 28 is 268,435,456
2 to the 29 is 536,870,912
2 to the 30 is 1,073,741,824
2 to the 31 is 2,147,483,648
2 to the 32 is 4,294,967,296
2 to the 33 is 8,589,934,592
2 to the 34 is 17,179,869,184
2 to the 35 is 34,359,738,368
2 to the 36 is 68,719,476,736
2 to the 37 is 137,438,953,472
2 to the 38 is 274,877,906,944
2 to the 39 is 549,755,813,888
2 to the 40 is 1,099,511,627,776
2 to the 41 is 2,199,023,255,552
2 to the 42 is 4,398,046,511,104
2 to the 43 is 8,796,093,022,208
2 to the 44 is 17,592,186,044,416
2 to the 45 is 35,184,372,088,832
2 to the 46 is 70,368,744,177,664
2 to the 47 is 140,737,488,355,328
2 to the 48 is 281,474,976,710,656
2 to the 49 is 562,949,953,421,312
2 to the 50 is 1,125,899,906,842,624

2 to the 51 is 2,251,799,813,685,248
2 to the 52 is 4,503,599,627,370,496
2 to the 53 is 9,007,199,254,740,992
2 to the 54 is 18,014,398,509,481,984
2 to the 55 is 36,028,797,018,963,968
2 to the 56 is 72,057,594,037,927,936
2 to the 57 is 144,115,188,075,855,872
2 to the 58 is 288,230,376,151,711,744
2 to the 59 is 576,460,752,303,423,488
2 to the 60 is 1,152,921,504,606,846,976
2 to the 61 is 2,305,843,009,213,693,952
2 to the 62 is 4,611,686,018,427,387,904
2 to the 63 is 9,223,372,036,854,775,808
2 to the 64 is 18,446,744,073,709,551,616
2 to the 65 is 36,893,488,147,419,103,232
2 to the 66 is 73,786,976,294,838,206,464
2 to the 67 is 147,573,952,589,676,412,928
2 to the 68 is 295,147,905,179,352,825,856
2 to the 69 is 590,295,810,358,705,651,712
2 to the 70 is 1,180,591,620,717,411,303,424
2 to the 71 is 2,361,183,241,434,822,606,848
2 to the 72 is 4,722,366,482,869,645,213,696
2 to the 73 is 9,444,732,965,739,290,427,392
2 to the 74 is 18,889,465,931,478,580,854,784
2 to the 75 is 37,778,931,862,957,161,709,568
2 to the 76 is 75,557,863,725,914,323,419,136
2 to the 77 is 151,115,727,451,828,646,838,272
2 to the 78 is 302,231,454,903,657,293,676,544
2 to the 79 is 604,462,909,807,314,587,353,088
2 to the 80 is 1,208,925,819,614,629,174,706,176
2 to the 81 is 2,417,851,639,229,258,349,412,352
2 to the 82 is 4,835,703,278,458,516,698,824,704
2 to the 83 is 9,671,406,556,917,033,397,649,408
2 to the 84 is 19,342,813,113,834,066,795,298,816
2 to the 85 is 38,685,626,227,668,133,590,597,632
2 to the 86 is 77,371,252,455,336,267,181,195,264
2 to the 87 is 154,742,504,910,672,534,362,390,528
2 to the 88 is 309,485,009,821,345,068,724,781,056
2 to the 89 is 618,970,019,642,690,137,449,562,112
2 to the 90 is 1,237,940,039,285,380,274,899,124,224
2 to the 91 is 2,475,880,078,570,760,549,798,248,448
2 to the 92 is 4,951,760,157,141,521,099,596,496,896
2 to the 93 is 9,903,520,314,283,042,199,192,993,792
2 to the 94 is 19,807,040,628,566,084,398,385,987,584
2 to the 95 is 39,614,081,257,132,168,796,771,975,168
2 to the 96 is 79,228,162,514,264,337,593,543,950,336
2 to the 97 is 158,456,325,028,528,675,187,087,900,672
2 to the 98 is 316,912,650,057,057,350,374,175,801,344
2 to the 99 is 633,825,300,114,114,700,748,351,602,688
2 to the 100 is 1,267,650,600,228,229,401,496,703,205,376

Appendix

# Prime Numbers up to 2000 (in groups of 250)

| | | | | | | | |
|---|---|---|---|---|---|---|---|
| 2 | 251 | 503 | 751 | 1009 | 1259 | 1511 | 1753 |
| 3 | 257 | 509 | 757 | 1013 | 1277 | 1523 | 1759 |
| 5 | 263 | 521 | 761 | 1019 | 1279 | 1531 | 1777 |
| 7 | 269 | 523 | 769 | 1021 | 1283 | 1543 | 1783 |
| 11 | 271 | 541 | 773 | 1031 | 1289 | 1549 | 1787 |
| 13 | 277 | 547 | 787 | 1033 | 1291 | 1553 | 1789 |
| 17 | 281 | 557 | 797 | 1039 | 1297 | 1559 | 1801 |
| 19 | 283 | 563 | 809 | 1049 | 1301 | 1567 | 1811 |
| 23 | 293 | 569 | 811 | 1051 | 1303 | 1571 | 1823 |
| 29 | 307 | 571 | 821 | 1061 | 1307 | 1579 | 1831 |
| 31 | 311 | 577 | 823 | 1063 | 1319 | 1583 | 1847 |
| 37 | 313 | 587 | 827 | 1069 | 1321 | 1597 | 1861 |
| 41 | 317 | 593 | 829 | 1087 | 1327 | 1601 | 1867 |
| 43 | 331 | 599 | 839 | 1091 | 1361 | 1607 | 1871 |
| 47 | 337 | 601 | 853 | 1093 | 1367 | 1609 | 1873 |
| 53 | 347 | 607 | 857 | 1097 | 1373 | 1613 | 1877 |
| 59 | 349 | 613 | 859 | 1103 | 1381 | 1619 | 1879 |
| 61 | 353 | 617 | 863 | 1109 | 1399 | 1621 | 1889 |
| 67 | 359 | 619 | 877 | 1117 | 1409 | 1627 | 1901 |
| 71 | 367 | 631 | 881 | 1123 | 1423 | 1637 | 1907 |
| 73 | 373 | 641 | 883 | 1129 | 1427 | 1657 | 1913 |
| 79 | 379 | 643 | 887 | 1151 | 1429 | 1663 | 1931 |
| 83 | 383 | 647 | 907 | 1153 | 1433 | 1667 | 1933 |
| 89 | 389 | 653 | 911 | 1163 | 1439 | 1669 | 1949 |
| 97 | 397 | 659 | 919 | 1171 | 1447 | 1693 | 1951 |
| 101 | 401 | 661 | 929 | 1181 | 1451 | 1697 | 1973 |
| 103 | 409 | 673 | 937 | 1187 | 1453 | 1699 | 1979 |
| 107 | 419 | 677 | 941 | 1193 | 1459 | 1709 | 1987 |
| 109 | 421 | 683 | 947 | 1201 | 1471 | 1721 | 1993 |
| 113 | 431 | 691 | 953 | 1213 | 1481 | 1723 | 1997 |
| 127 | 433 | 701 | 967 | 1217 | 1483 | 1733 | 1999 |
| 131 | 439 | 709 | 971 | 1223 | 1487 | 1741 | |
| 137 | 443 | 719 | 977 | 1229 | 1489 | 1747 | |
| 139 | 449 | 727 | 983 | 1231 | 1493 | | |
| 149 | 457 | 733 | 991 | 1237 | 1499 | | |
| 151 | 461 | 739 | 997 | 1249 | | | |
| 157 | 463 | 743 | | | | | |
| 163 | 467 | | | | | | |
| 167 | 479 | | | | | | |
| 173 | 487 | | | | | | |
| 179 | 491 | | | | | | |
| 181 | 499 | | | | | | |
| 191 | | | | | | | |
| 193 | | | | | | | |
| 197 | | | | | | | |
| 199 | | | | | | | |
| 211 | | | | | | | |
| 223 | | | | | | | |
| 227 | | | | | | | |
| 229 | | | | | | | |
| 233 | | | | | | | |
| 239 | | | | | | | |
| 241 | | | | | | | |

Appendix

# Even Numbers as the Sum of Two Primes

**4** = 2+2
**6** = 3+3
**8** = 3+5
**10** = 3+7; 5+5
**12** = 5+7
**14** = 3+11; 7+7
**16** = 3+13; 5+11
**18** = 5+13; 7+11
**20** = 3+17; 7+13
**22** = 3+19; 5+17; 11+11
**24** = 5+19; 7+17; 11+13
**26** = 3+23; 7+19; 13+13
**28** = 5+23; 11+17
**30** = 7+23; 11+19; 13+17
**32** = 3+29; 13+19
**34** = 3+31; 5+29; 11+23; 17+17
**36** = 5+31; 7+29; 13+23; 17+19
**38** = 7+31; 19+19
**40** = 3+37; 11+29; 17+23
**42** = 5+37; 11+31; 13+29; 19+23
**44** = 3+41; 7+37; 13+31
**46** = 3+43; 5+41; 17+29; 23+23
**48** = 5+43; 7+41; 11+37; 17+31; 19+29
**50** = 3+47; 7+43; 13+37; 19+31
**52** = 5+47; 11+41; 23+29

**54** = 7+47; 11+43; 13+41; 17+37; 23+31
**56** = 3+53; 13+43; 19+37
**58** = 5+53; 11+47; 17+41; 29+29
**60** = 7+53; 13+47; 17+43; 19+41; 23+37; 29+31
**62** = 3+59; 19+43; 31+31
**64** = 3+61; 5+59; 11+53; 17+47; 23+41
**66** = 5+61; 7+59; 13+53; 19+47; 23+43; 29+37
**68** = 7+61; 31+37
**70** = 3+67; 11+59; 17+53; 23+47; 29+41
**72** = 5+67; 11+61; 13+59; 19+53; 29+43; 31+41
**74** = 3+71; 7+67; 13+61; 31+43; 37+37
**76** = 3+73; 5+71; 17+59; 23+53; 29+47
**78** = 5+73; 7+71; 11+67; 17+61; 19+59; 31+47; 37+41
**80** = 7+73; 13+67; 19+61; 37+43
**82** = 3+79; 11+71; 23+59; 29+53; 41+41
**84** = 5+79; 11+73; 13+71; 17+67; 23+61; 31+53; 37+47; 41+43
**86** = 3+83; 7+79; 13+73; 19+67; 43+43
**88** = 5+83; 17+71; 29+59; 41+47
**90** = 7+83; 11+79; 17+73; 19+71; 23+67; 29+61; 31+59; 37+53; 43+47
**92** = 3+89; 13+79; 19+73; 31+61
**94** = 5+89; 11+83; 23+71; 41+53; 47+47
**96** = 7+89; 13+83; 17+79; 23+73; 29+67; 37+59; 43+53
**98** = 19+79; 31+67; 37+61
**100** = 3+97; 11+89; 17+83; 29+71; 41+59; 47+53

**102** = 5+97; 13+89; 19+83; 23+79; 29+73; 31+71; 41+61; 43+59
**104** = 3+101; 7+97; 31+73; 37+67; 43+61
**106** = 3+103; 5+101; 17+89; 23+83; 47+59; 53+53
**108** = 5+103; 7+101; 11+97; 19+89; 29+79; 37+71; 41+67; 47+61
**110** = 3+107; 7+103; 13+97; 31+79; 37+73; 43+67
**112** = 3+109; 5+107; 11+101; 23+89; 29+83; 41+71; 53+59
**114** = 5+109; 7+107; 11+103; 13+101; 17+97; 31+83; 41+73; 43+71; 47+67; 53+61
**116** = 3+113; 7+109; 13+103; 19+97; 37+79; 43+73
**118** = 5+113; 11+107; 17+101; 29+89; 47+71; 59+59
**120** = 7+113; 11+109; 13+107; 17+103; 19+101; 23+97; 31+89; 37+83; 41+79; 47+73; 53+67; 59+61
**122** = 13+109; 19+103; 43+79; 61+61
**124** = 11+113; 17+107; 23+101; 41+83; 53+71
**126** = 13+113; 17+109; 19+107; 23+103; 29+97; 37+89; 43+83; 47+79; 53+73; 59+67
**128** = 19+109; 31+97; 61+67
**130** = 3+127; 17+113; 23+107; 29+101; 41+89; 47+83; 59+71
**132** = 5+127; 19+113; 23+109; 29+103; 31+101; 43+89; 53+79; 59+73; 61+71
**134** = 3+131; 7+127; 31+103; 37+97; 61+73; 67+67
**136** = 5+131; 23+113; 29+107; 47+89; 53+83
**138** = 7+131; 11+127; 29+109; 31+107; 37+101; 41+97; 59+79; 67+71
**140** = 3+137; 13+127; 31+109; 37+103; 43+97; 61+79; 67+73
**142** = 3+139; 5+137; 11+131; 29+113; 41+101; 53+89; 59+83; 71+71
**144** = 5+139; 7+137; 13+131; 17+127; 31+113; 37+107; 41+103; 43+101; 47+97; 61+83; 71+73
**146** = 7+139; 19+127; 37+109; 43+103; 67+79; 73+73
**148** = 11+137; 17+131; 41+107; 47+101; 59+89
**150** = 11+139; 13+137; 19+131; 23+127; 37+113; 41+109; 43+107; 47+103; 53+97; 61+89; 67+83; 71+79

Appendix
# Odd Numbers as the Difference of Two Squares

**3** = 4 - 1
**5** = 9 - 4
**7** = 16 - 9
**9** = 25 - 16
**11** = 36 - 25
**13** = 49 - 36
**15** = 16 - 1 ; 64 - 49
**17** = 81 - 64
**19** = 100 - 81
**21** = 25 - 4 ; 121 - 100
**23** = 144 - 121
**25** = 169 - 144
**27** = 36 - 9 ; 196 - 169
**29** = 225 - 196
**31** = 256 - 225
**33** = 49 - 16 ; 289 - 256
**35** = 36 - 1 ; 324 - 289
**37** = 361 - 324
**39** = 64 - 25 ; 400 - 361
**41** = 441 - 400
**43** = 484 - 441
**45** = 49 - 4 ; 81 - 36 ; 529 - 484
**47** = 576 - 529
**49** = 625 - 576
**51** = 100 - 49 ; 676 - 625
**53** = 729 - 676
**55** = 64 - 9 ; 784 - 729
**57** = 121 - 64 ; 841 - 784
**59** = 900 - 841
**61** = 961 - 900
**63** = 64 - 1 ; 144 - 81 ; 1024 - 961
**65** = 81 - 16 ; 1089 - 1024
**67** = 1156 - 1089
**69** = 169 - 100 ; 1225 - 1156
**71** = 1296 - 1225
**73** = 1369 - 1296
**75** = 100 - 25 ; 196 - 121 ; 1444 - 1369
**77** = 81 - 4 ; 1521 - 1444
**79** = 1600 - 1521
**81** = 225 - 144 ; 1681 - 1600
**83** = 1764 - 1681
**85** = 121 - 36 ; 1849 - 1764
**87** = 256 - 169 ; 1936 - 1849
**89** = 2025 - 1936
**91** = 100 - 9 ; 2116 - 2025
**93** = 289 - 196 ; 2209 - 2116
**95** = 144 - 49 ; 2304 - 2209
**97** = 2401 - 2304
**99** = 100 - 1 ; 324 - 225 ; 2500 - 2401

**101** = 2601 - 2500
**103** = 2704 - 2601
**105** = 121 - 16 ; 169 - 64 ; 361 - 256 ; 2809 - 2704
**107** = 2916 - 2809
**109** = 3025 - 2916
**111** = 400 - 289 ; 3136 - 3025
**113** = 3249 - 3136
**115** = 196 - 81 ; 3364 - 3249
**117** = 121 - 4 ; 441 - 324 ; 3481 - 3364
**119** = 144 - 25 ; 3600 - 3481
**121** = 3721 - 3600
**123** = 484 - 361 ; 3844 - 3721
**125** = 225 - 100 ; 3969 - 3844
**127** = 4096 - 3969
**129** = 529 - 400 ; 4225 - 4096
**131** = 4356 - 4225
**133** = 169 - 36 ; 4489 - 4356
**135** = 144 - 9 ; 256 - 121 ; 576 - 441 ; 4624 - 4489
**137** = 4761 - 4624
**139** = 4900 - 4761
**141** = 625 - 484 ; 5041 - 4900
**143** = 144 - 1 ; 5184 - 5041
**145** = 289 - 144 ; 5329 - 5184
**147** = 196 - 49 ; 676 - 529 ; 5476 - 5329
**149** = 5625 - 5476
**151** = 5776 - 5625
**153** = 169 - 16 ; 729 - 576 ; 5929 - 5776
**155** = 324 - 169 ; 6084 - 5929
**157** = 6241 - 6084
**159** = 784 - 625 ; 6400 - 6241
**161** = 225 - 64 ; 6561 - 6400
**163** = 6724 - 6561
**165** = 169 - 4 ; 361 - 196 ; 841 - 676 ; 6889 - 6724
**167** = 7056 - 6889
**169** = 7225 - 7056
**171** = 196 - 25 ; 900 - 729 ; 7396 - 7225
**173** = 7569 - 7396
**175** = 256 - 81 ; 400 - 225 ; 7744 - 7569
**177** = 961 - 784 ; 7921 - 7744
**179** = 8100 - 7921
**181** = 8281 - 8100
**183** = 1024 - 841 ; 8464 - 8281
**185** = 441 - 256 ; 8649 - 8464
**187** = 196 - 9 ; 8836 - 8649
**189** = 225 - 36 ; 289 - 100 ; 1089 - 900 ; 9025 - 8836
**191** = 9216 - 9025
**193** = 9409 - 9216
**195** = 196 - 1 ; 484 - 289 ; 1156 - 961 ; 9604 - 9409
**197** = 9801 - 9604
**199** = 10000 - 9801

Appendix

# Numbers as the Sum of Two Squares

(Numbers that are missing cannot be expressed as the sum of two squares.)

**2** = 1 + 1
**5** = 1 + 4
**8** = 4 + 4
**10** = 1 + 9
**13** = 4 + 9
**17** = 1 + 16
**18** = 9 + 9
**20** = 4 + 16
**25** = 9 + 16
**26** = 1 + 25
**29** = 4 + 25
**32** = 16 + 16
**34** = 9 + 25
**37** = 1 + 36
**40** = 4 + 36
**41** = 16 + 25
**45** = 9 + 36
**50** = 1 + 49 ; 25 + 25
**52** = 16 + 36
**53** = 4 + 49
**58** = 9 + 49
**61** = 25 + 36
**65** = 1 + 64 ; 16 + 49
**68** = 4 + 64
**72** = 36 + 36
**73** = 9 + 64
**74** = 25 + 49
**80** = 16 + 64
**82** = 1 + 81
**85** = 4 + 81 ; 36 + 49
**89** = 25 + 64
**90** = 9 + 81
**97** = 16 + 81
**98** = 49 + 49
**100** = 36 + 64
**101** = 1 + 100
**104** = 4 + 100
**106** = 25 + 81
**109** = 9 + 100
**113** = 49 + 64
**116** = 16 + 100
**117** = 36 + 81
**122** = 1 + 121
**125** = 4 + 121 ; 25 + 100
**128** = 64 + 64
**130** = 9 + 121 ; 49 + 81
**136** = 36 + 100
**137** = 16 + 121
**145** = 1 + 144 ; 64 + 81
**146** = 25 + 121

**148** = 4 + 144
**149** = 49 + 100
**153** = 9 + 144
**157** = 36 + 121
**160** = 16 + 144
**162** = 81 + 81
**164** = 64 + 100
**169** = 25 + 144
**170** = 1 + 169 ; 49 + 121
**173** = 4 + 169
**178** = 9 + 169
**180** = 36 + 144
**181** = 81 + 100
**185** = 16 + 169 ; 64 + 121
**193** = 49 + 144
**194** = 25 + 169
**197** = 1 + 196
**200** = 4 + 196 ; 100 + 100
**202** = 81 + 121
**205** = 9 + 196 ; 36 + 169
**208** = 64 + 144
**212** = 16 + 196
**218** = 49 + 169
**221** = 25 + 196 ; 100 + 121
**225** = 81 + 144
**226** = 1 + 225
**229** = 4 + 225
**232** = 36 + 196
**233** = 64 + 169
**234** = 9 + 225
**241** = 16 + 225
**242** = 121 + 121
**244** = 100 + 144
**245** = 49 + 196
**250** = 25 + 225 ; 81 + 169
**257** = 1 + 256
**260** = 4 + 256 ; 64 + 196
**261** = 36 + 225
**265** = 9 + 256 ; 121 + 144
**269** = 100 + 169
**272** = 16 + 256
**274** = 49 + 225
**277** = 81 + 196
**281** = 25 + 256
**288** = 144 + 144
**289** = 64 + 225
**290** = 1 + 289 ; 121 + 169
**292** = 36 + 256
**293** = 4 + 289
**296** = 100 + 196

**298** = 9 + 289
**305** = 16 + 289 ; 49 + 256
**306** = 81 + 225
**313** = 144 + 169
**314** = 25 + 289
**317** = 121 + 196
**320** = 64 + 256
**325** = 1 + 324 ; 36 + 289 ; 100 + 225
**328** = 4 + 324
**333** = 9 + 324
**337** = 81 + 256
**338** = 49 + 289 ; 169 + 169
**340** = 16 + 324 ; 144 + 196
**346** = 121 + 225
**349** = 25 + 324
**353** = 64 + 289
**356** = 100 + 256
**360** = 36 + 324
**362** = 1 + 361
**365** = 4 + 361 ; 169 + 196
**369** = 144 + 225
**370** = 9 + 361 ; 81 + 289
**373** = 49 + 324
**377** = 16 + 361 ; 121 + 256
**386** = 25 + 361
**388** = 64 + 324
**389** = 100 + 289
**392** = 196 + 196
**394** = 169 + 225
**397** = 36 + 361
**400** = 144 + 256
**401** = 1 + 400
**404** = 4 + 400
**405** = 81 + 324
**409** = 9 + 400
**410** = 49 + 361 ; 121 + 289
**416** = 16 + 400
**421** = 196 + 225
**424** = 100 + 324
**425** = 25 + 400 ; 64 + 361 ; 169 + 256
**433** = 144 + 289
**436** = 36 + 400
**442** = 1 + 441 ; 81 + 361

*The first number that can be expressed in 4 ways is...*
**1105** = 16 + 1089; 81 + 1024;
144 + 961; 529 + 576

108